Liudmyla Popova
Irina Vasylyeva

Neurohumoral Status and Aggression

Liudmyla Popova
Irina Vasylyeva

Neurohumoral Status and Aggression

The Role of Neurohumoral Status Features in the Development of Different Types of Aggression

LAP LAMBERT Academic Publishing

Impressum / Imprint

Bibliografische Information der Deutschen Nationalbibliothek: Die Deutsche Nationalbibliothek verzeichnet diese Publikation in der Deutschen Nationalbibliografie; detaillierte bibliografische Daten sind im Internet über http://dnb.d-nb.de abrufbar.

Alle in diesem Buch genannten Marken und Produktnamen unterliegen warenzeichen-, marken- oder patentrechtlichem Schutz bzw. sind Warenzeichen oder eingetragene Warenzeichen der jeweiligen Inhaber. Die Wiedergabe von Marken, Produktnamen, Gebrauchsnamen, Handelsnamen, Warenbezeichnungen u.s.w. in diesem Werk berechtigt auch ohne besondere Kennzeichnung nicht zu der Annahme, dass solche Namen im Sinne der Warenzeichen- und Markenschutzgesetzgebung als frei zu betrachten wären und daher von jedermann benutzt werden dürften.

Bibliographic information published by the Deutsche Nationalbibliothek: The Deutsche Nationalbibliothek lists this publication in the Deutsche Nationalbibliografie; detailed bibliographic data are available in the Internet at http://dnb.d-nb.de.

Any brand names and product names mentioned in this book are subject to trademark, brand or patent protection and are trademarks or registered trademarks of their respective holders. The use of brand names, product names, common names, trade names, product descriptions etc. even without a particular marking in this work is in no way to be construed to mean that such names may be regarded as unrestricted in respect of trademark and brand protection legislation and could thus be used by anyone.

Coverbild / Cover image: www.ingimage.com

Verlag / Publisher:
LAP LAMBERT Academic Publishing
ist ein Imprint der / is a trademark of
OmniScriptum GmbH & Co. KG
Heinrich-Böcking-Str. 6-8, 66121 Saarbrücken, Deutschland / Germany
Email: info@lap-publishing.com

Herstellung: siehe letzte Seite /
Printed at: see last page
ISBN: 978-3-659-62037-9

CONTENT

ABBREVIATIONS

ACTH	adrenocorticotropic hormone
AR	androgen receptors
AVP	arginine-vasopressin
BLA	basolateral amygdala
BNST	bed nucleus of stria terminalis
CBG	corticosteroid binding protein
CRH	corticotropin-releasing hormone
CRHR	corticotropin-releasing hormone
DA	dopamine
DHT	dihydrotestosterone
ER	estrogen receptors
GABA	gamma-aminobutyric acid
5-HIAA	5-hydroxyindoleacetic acid
HPA-axis	hypothalamic pituitary adrenal axis
5-HT	serotonin
KO	knockout
MAO	monoamine oxidase
LC	locus coeruleus
NA	noradrenaline
PACAP	pituitay adenylate cyclase-activating polypeptide
PAG	periaqueductal gray
PFC	prefrontal cortex
PVN	paraventricular nucleus
T_3	triiodothyronine
T_4	thyroxine
VTA	ventral tegmental area

INTRODUCTION

Aggression is significant problem of modern society. Nowadays, more and more people around the world suffer from aggression on the street, at home, in the family, from the terrorists, of local wars. Every year more than 700.000 people worldwide die because of assault.

The studies of genetically determined characteristics of neurohumoral status in animals inclined to aggressive or submissive type of behavior, as well as identification of their specific markers are essential to prevent depressive and aggressive states in humans. Therefore, great importance is the study of neurobiological mechanisms of aggression and depression in humans and studies of genetically determined characteristics of neurohumoral status in animals predisposed to dominant or submissive type of behavior. A definition of peripheral markers of these features is important to prevent the development of such conditions.

Nowadays, the most attention in the mechanisms of development of depressive and aggressive states was given to monoaminergic systems of the brain, but literature data on the role of serotonin, norepinephrine and dopamine in the development and regulation of aggression and depression are rather contradictory [8, 29, 198].

It was shown that the expression of aggressive behavior in adult is provided by male sex steroids [37] which in the perinatal period are involved in the formation of neuronal networks [209], and in adult testosterone contributes to the modulation of neuronal pathways that regulate aggression [132]. However, it was not investigated which mediator systems are involved in this process.

There are literature data that androgens play a critical role in the regulation of the hypothalamic pituitary adrenal (HPA) axis [110] and HHA axis dysregulation is often associated with anxiety and depression, as well as violence and aggression [41]. It is known that the hippocampus, a key structure of emotiogenic limbic-neocortical system of the brain, is involved in the mediation of the response to stress and controls the duration of this response [78], and frontal cortex has an inhibiting effect on aggression and violence [52].

However, the nature of relationships between sex hormones, HPA axis components and monoaminergic neurotransmission in hippocampus and frontal cortex in the mechanisms of the trigger, development and control of aggressive behavior was insufficantly studied.

Neurophysiological and biochemical investigations of neuromediatory status in aggression or depression concerned mainly the regions of emotiogenic limbicocortical system of brain but cerebellar implication in aggression and depression development was not investigated. Cerebellum is important region in regulation of motor function and is obviously involved in aggressive behavior realization. Moreover according to recent advances in cerebellar research it participates in regulation of the highest cognitive [179] and psychoemotional functions [87].

In this regard, we have investigated the neurohumoral status features in rats with dominant and submissive types of behavior. To achieve this purpose the following objectives have been defined:

- To investigate the hormonal status of rats with alternative types of behavior:

a) to determine testosterone and 17β-estradiol levels in the blood plasma of rats with alternative types of behavior;

b) to determine the contents of the hypothalamic pituitary adrenal axis hormones (adrenocorticotropic hormone and corticosterone) in the blood plasma of rats with alternative types of behavior;

c) to determine the content of thyroid hormones (T_3 and T_4) in blood plasma of rats with alternative types of behavior.

- To determine the content of monoamines (serotonin, noradrenaline, dopamine) in frontal cortex, hippocampus and cerebellum of rats with alternative types of behavior.

- To identify the relationship between the types of behavior and the nature of the relationship between hormones and components of monoaminergic system.

Section 1

NEURO HUMORAL MECHANISMS OF AGGRESSION DEVELOPMENT (short review of literature data)

Two main forms of human excessive/violent aggression are described, namely impulsive-reactive-hostile-affective aggression and controlled-proactive-instrumental-predatory aggression [204]. The first form of aggression is seen in patients with depression, PTSD, or intermittent explosive disorder. The second type of aggression can be revealed in patients with personality disorders (conduct, antisocial, and borderline disorder), but may also be found in individuals without noticeable emotional or social deficits [132]. The first type of aggression is characterized by high emotional reactions and autonomic responses including high glucocorticoid levels. The second type is associated with low emotional, autonomic, and glucocorticoid responses [71].

Both the neuroendocrine system and the brain mechanisms underlying gender-specific behavior are known to be organized by steroid sex hormones during specific sensitive phases of early fetal and perinatal development [85]. According to the classic view of sexual differentiation in mammalian species the sex differences in the brain structure and function are programmed by exposure to testosterone produced by the fetal testis acting during a critical period in perinatal development [38].

Testosterone synthesized by fetal testis diffuses into the male brain where it is locally aromatized to estradiol and then initiates the process of masculinization [161]. 17-β-Estradiol is formed more actively in the male mouse brain than in the female during both the prenatal and the neonatal periods. These sex differences are regionally specific in the brain. [85].

Aromatase, the key enzyme in the conversion of androgens to estrogens, controls many physiological and behavioral processes. In vertebrates, the regulation of aromatase expression in the brain has been implicated in the modulation of male sexual and aggressive behaviors [61]. The local synthesis of estrogen in the brain is a dynamic and regulated process that varies with age, sex, and physiologic status. Aromatase expression in the preoptic area/hypothalamus declines to low levels after birth and into adulthood [161]. In contrast to aromatase expression in the preoptic area and hypothalamus, gonadal steroids do not regulate aromatase in most other brain areas, including the amygdala and hippocampus [2]. The conversion of androgens into estrogens in the brain is a key mechanism by which testosterone regulates many physiological and behavioral processes throughout an animal's life

[161]. However the maintenance of normal synaptic density in hippocampus is androgen-dependent, via a mechanism that does not require intermediate estrogen biosynthesis. [113].

In perinatal period androgens facilitate the formation of neuronal networks which are expressed by aggressive behaviour in adults [37]. In the mutant mouse males selectively lacking AR expression in the nervous system, urogenital tract was normally developed, but sexual motivation and aggressive behaviors were affected. These alterations occurred despite increased levels of testosterone and its metabolites, and an unaffected number of immunoreactive cells to estradiol [153].

Masculinization of brain is evidenced in adults by a capacity to express male – typical sexual behaviour and high level of aggression [161]. It is believed that the acute rise in testosterone contributes to the modulation of neuronal networks regulating aggression [44].

In male rodents the removal of androgen endogenous sources by gonadectomy causes increased anxiety and depressive behavior that is normalized by systemic administration of testosterone [57]. Similarly, in men the aging process is accompanied by the decrease of androgen levels tha can lead to the development of symptoms overlaping with depression symptoms [7].

Organizing effects during fetal life as well as activating effects of sex hormones on the HPA axis have been reported [186]. Findings obtained by Bingham B. et al. (2011) underscore an organizing influence of both androgen receptors and androgen conversion to estrogen on HPA habituation to repeated psychogenic stress, which appears to occur independently of the activational effects of testosterone [17]. Significantly greater blood plasma ACTH and corticosterone levels were observed in gonadectomized male rats versus intacts after exposure to stress. Testosterone or dihydrotestosterone propionate treatment of gonadectomized animals returned poststress levels of these parameters to intact levels [74].

HPA axis is the major neuroendocrine system that responds to stress. HPA-axis activation in response to physical and emotional stress is accompanied by a number of neuronal and emotional responses aimed at maintaining homeostasis [41]. Activity HPA axis is controlled by a subset of neurons in the parvo cellular part of paraventricular nucleus (PVN) of the hypothalamus that secrete corticotropin-releasing hormone (CRH) and arginine-vasopressin (AVP) into the pituitary portal system. These neuroendocrine neurons receive afferent fibers from the brain stem, and regions of the anterior and limbic brain. In the adenohypophysis CRH and somewhat less AVP stimulate the synthesis and secretion of adrenocorticotropic hormone (ACTH), which in turn stimulates the production of glucocorticoids (cortisol – in humans, corticosterone - in rodents) by adrenal cortex. Thus, rapid activation of HPA axis in response to stress is directed by hypothalamic CRH.

Abnormalities in the regulation by CRH play an important role in the development of depression and anxiety [155]. CRH realizes its effects through receptors CRHR1 and CRHR2. Both CRHR1 and CRHR2 have integral role in the regulation of sensitivity to stress. Changes in receptor expression may be linked to disturbances in behavior [9]. For example, CRHR1 and CRHR2 knockout (KO) animals have hypo- and hypersensitivity to stress respectively [9]. Moreover, CRHR2-deficient male mice exhibit enhanced anxious behaviour in several tests of anxiety in contrast to mice lacking CRHR1 [93]. Regulation of the relative contribution of the two CRH receptors to brain CRH pathways may be essential in coordinating physiological responses to stress [9].

Sex hormones play a key role in the regulation of CRH central expression [133]. Basal CRH expression is lower in males, than in females [152]. The ACTH response to conditioned stress was significantly potentiated in gonadectomized males compared with sham-operated and gonadectomized DHT-treated animals [15]. Long -term castration increases hypothalamic CRH content and CRH-IR cell numbers in the PVN by removal of an androgen-dependent repression [16]. The studies of M.J. Weiser et al. (2008) indicated that DHT significantly increased CRHR2 mRNA expression in the hippocampus, hypothalamus, and lateral septum. These changes in CRHR2 mRNA correspond with changes in CRHR2 binding within the lateral septum. It appears that this regulation is mediated specifically through androgen receptor activation, at least in the hippocampus [208].

Thus, on the one hand, androgens affect sensitivity to stress. On the other hand, HPA axis hormones play an important role in the mediation of prenatal stress induced disorders of brain development programming processes. In prenatally stressed males an attenuation of sex-specific pattern of the protein fractions (on the 5th postnatal day), steroid aromatase activity (on the 10th postnatal day) in the brain preoptic area, and a decrease of male copulatory behavior, hypothalamic noradrenaline and plasma corticosterone responses to an acute stress, an increase in HPA responses to noradrenergic stimulation and other effects in adulthood were observed [156].

The exposure to stress was shown to activate the neurons in lateral septum [34]. Caudal dorsal region of lateral septum receives dense inputs from the hippocampus, dorsal raphe, locus coeruleus and ventral tegmental area. All these areas synthesize monoamine [50]. Ventral region is the only septal area with bilateral ties with paraventricular nucleus and can therefore play an important role in regulating the activity of HPA axis [159].

The hippocampus plays an integral role in the regulation of the neuroendocrine response to stress, particularly to psychogenic stress [78]. The hippocampus is critical in mediation of glucocorticoid-dependent negative feedback and thus contributes to control of the duration of response to stress. Glutamatergic projections from the hippocampus reach GABA-ergic neurons in the amygdala, lateral septum, bed nucleus of the stria terminalis (BNST), prefrontal cortex (PFC). The projections of these neurons reach the paraventricular nucleus of the hypothalamus and change CRH and AVP secretion [78]. According to D.A. Bevzuk (2007), the hippocampus performs both inhibitory and excitatory effects on aggressive behavior initiated by trigger mechanisms of ventromedial hypothalamus [14].

Glucocorticoids influence the neurobehavioral functions in different brain areas [105]. Chronic corticosterone treatment is able to cause dendrites to retract in CA3 hippocampus. Chronic restraint stress is known to reduce dentate gyrus proliferation [121]. As HPA axis generates a return loop through some areas of the brain such as the hippocampus and amygdala, such changes in hippocampal neurons, caused by chronic stress, reduce its influence on the HPA axis, which leads to its hyperactivity [141].

Chronic stress reduced hippocampal neurogenesis [105]. The evidence of the strong relationship between depression and the hippocampus atrophy, neurogenesis inhibition and significant reduction of glial cells was found [149]. Brain imaging studies showed the reduced hippocampal volume in depressed patients [48]. Different animal models of depression are associated with reductions in hippocampal neurogenesis [104, 184].

Neocortex and hippocampus is possible to play a crucial role in the development of symptoms belonging to the cognitive deficits manifested in depressed patients [47]. Control of higher cognitive functions is localized in the frontal cortex. Thinning and decreased gray matter density of the insula disorganizes prefrontal circuits, diminishing the inhibitory influence of the prefrontal cortex on anger, aggression, cruelty, and impulsivity, and increasing a person's likelihood of aggressive behavior [52]. The amygdala is thought to generate emotional responses, and the prefrontal brain regions to regulate those responses [200]. It is believed that the frontal cortex, cingulate and olfactory cortex, nucleus accumbens are involved in the control of anxiety, mentale state, cognitive functions, emotions and behavior [54].

Monoaminergic brain systems have a great influence on the emotions and behavior in humans and animals [132, 198]. It is believed that monoaminergic neurotransmitter system provides the homeostasis both brain and whole organism [55]. The psycho-emotional state of the organism may be modified by changing the activity of noradrenergic and serotoninergic systems by chemicals [204].

Noradrenergic transmission is thout to be essential to maintain and improve vigilance, motivation and self-perception [27]. Noradrenaline (NA) was related to increased social engagement and cooperation and reduction in self-focus [193]. Reboxetine (new antidepressant with a selective effect on NA) increases social adaptation [194]. Serotonin (5-hydroxytryptamine, 5-HT) affects impulsivity and irritability, while dopamine (DA) is important in the regulation of drive [27].

It should also be noted that the literature on the role of monoamines in the development of aggression is rather contradictory [32, 132].

In human and rats serotonin regulates aggressive behavior by binding to 5-HT 1A and 1B receptors [45]. It is believed that changes in the development and function of key neuronal serotonin pathway between the dorsal raphe nucleus, medial prefrontal cortex, amygdla has a great influence on individual differences in response to stress and risk of stress-related diseases in humans [83].

In cerebrospinal fluid of humans low levels of serotonin and 5-hydroxyindolacetic acid (5-HIAA) were shown to correlate with high aggressiveness [132]. Another study revealed a positive correlation between aggressiveness and concentration serotonin, 5-HIAA, NA, DA and 3,4-dihydroxyphenylacetic acid [198].

Interesting results were obtained in studies of Neumann I.D. et al. (2010) [132]. They revealed a higher serotoninergic neurotransmission in rats with low anxiety-related behavior (LAB-rats) compared with rats with high anxiety-related behavior (HAB-rats). Attention is drawn to the fact that male LAB-rats and to a lesser extent male HAB-rats showed a high form of aggression. The authors suggest a different role of serotonin in the adaptive forms of aggression, such as social dominance (increased serotoninergic neurotransmission) versus excessive or abnormal forms of aggression (reduction of serotoninergic neurotransmission).

Although other neurotransmitters are indirectly involved in depression, there is much evidence of reduced serotoninergic neurotransmission as a primary defect in depression [1]. It is believed that serotonin contributes significantly to the mechanisms of genetically determined individual differences in aggressiveness. Genes that encode key enzymes of its metabolism in the brain (tryptophan

11

hydroxylase and monoamine oxidase A-MAO-A) and 5-HT 1A receptors belong to a set of genes that modulate aggressive behavior [148]. Various studies have shown that serotonin regulates impulsivity, providing an inhibitory control of aggression [212]. According to other works, in contrast, increased serotonin transmission contributes to aggressiveness. For example, de Boer and Koolhaas [39] have shown that the compound S-15535, which acts primarily as an agonist of somato-dendrite 5-HT 1A autoreceptors and an antagonist of post-synaptic 5-HT 1A receptors, is very effective in reducing aggressive behavior. These data clearly show that specific antiaggressive effects of the agonists of 5-HT 1A and 5-HT 1B receptors are provided by a decrease rather than increase in serotonin neurotransmission. This assumption is consistent with the findings about the number of 5-HT 1A receptors in the cortico-limbic and the midbrain structures by the method of positron emission tomography, using radioligand. In individuals with greater aggressiveness, an increased number (or activity) of 5-HT 1A receptors has been found in the frontal cortex [212]. Obviously, they are postsynaptic, because 5-HT 1A-autoreceptors are characterized by somato-dendritic localization.

Dopamine is the predominant catecholamine neurotransmitter in the central nervous system of human. Dopamine receptors are mainly localized in the striatum, limbic system, cortex and infundibulum [145]. Dopaminergic system plays a major role in the physiology of the HPA axis regulation [145].

As with serotonin, there are clinical and preclinical data showing decreased as well as increased metabolism of DA in depression [8]. According to Oquendo M.A. et al. (2000), suicidal behaviors and the lethality of suicide attempts may be linked to the abnormalities in neurotransmitter systems similar to those found in patients with impulsive aggression, namely, lowered 5-HT transmission and enhanced DA and NA functioning [138].

Serotoninergic system provides physiological and tonic effects on the dopaminergic system through mesocorticolimbic 5-HT 2C receptors [46].

Almost all brain noradrenergic fibers arise in brainstem nuclei designated A1-A7. Approximately half of them belong to the locus coeruleus (LC) [150].

Neurons localized in LC have projections to the cerebral cortex and subcortical many areas, including the hippocampus, amygdala, thalamus and hypothalamus. These features of noradrenergic neuronal system determine its adaptability to rapid and global modulation of brain function in response to environmental change, as occurs under stress [26].

There is considerable evidence of the relationship between NA-ergic system of the brain and behavior associated with stress and anxiety. Exposure to stress is associated with an increase in firing of LC and with increased release and turnover of norepinephrine in brain regions which receive noradrenergic innervation. Increased activity of LC is also associated with behavioral manifestations of fear [26].

There is much evidence in violation of noradrenaline and serotonin systems neurotransmission in depression and anxiety states [154]. Many studies have been carried out on knockout animals. In dopamine-β-hydroxylase knockout mice, it was shown that the lack of NA did not greatly affect on anxiety in mice, but intact NA-transmission is necessary for the formation of some types of social memory and aggression (resident-intruder aggression) [115]. Data, obtained on the noradrenaline transporter knockout mice, suggest that enhanced noradrenergic function does not prevent situation –specific social learning but impedes the generalization of depression to heterotypic circumstances [72]. MAO-A knockout mice were shown to have the increased serotonin, noradrenaline and dopamine levels and to demonstrate aggressive behavior [83]. In other study it was shown that mice, lacking axons originating from the LC-NA neurons, presented increased anxiety-like behavior. Mice, lacking NA-ergic neurons in LC, showed not only increasd anxiety –like behavior but also increased depression-like behavior [88].

The findings obtained by Tanaka M. et al. (2000) suggest that the increased release of noradrenaline in the hypothalamus, amygdala and locus coeruleus is, in part, involved in the provocation of anxiety and/or fear in animals exposed to stress, and that the attenuation of this increase by benzodiazepine anxiolytics acting via the benzodiazepine receptor/GABAA receptor/chloride ionophore supramolecular complex may be the basic mechanism of action of these anxiolytic drug [188].

Discussing noradrenaline role in the mediation of fear and aggression, we should not forget the sympathetic nervous system and hormones of the adrenal medulla. Glucocorticoids and catecholamines are important stress hormones. Glucocorticoids of adrenal cortex stimulate phenylethanolamine-N-methyltransferase of adrenal medulla which converts noradrenaline into adrenaline [81, 171]. Classic anatomical studies have shown the existence of two different populations of chromaffin cells (adrenaline- and noradrenaline-containing) [40]. Glucocorticoids stimulate the activity phenylethanolamine-N-methyltransferase only in a subpopulation of adrenal-containing chromaffin cells [40].

Interestingly, the secretory response of adrenal medulla not only reflects the increasing fixed ratioA/NA what would be expected in the case of proportional release of NA and A. Instead, the ratio A/NA varies depending on the the magnitude and type of stimulus that triggers neuronal activation of the adrenal medulla [205]. This variability suggests the existence of selective control of two cell types. The existence of certain neuronal circuits between centers of the brain and chromaffin cells indicates the possibility of selective control [126].

Acetylcholine is the primary neurotransmitter mediating catecholamine secretion from the adrenal medulla [73]. Different firing patterns of splanchnic nerves and nicotinic or muscarinic receptors cause the selective release of noradrenaline or adrenaline, to adapt the body to the 'fight or flight' reaction [40]. Norepinephrine is called "lion hormone" and adrenaline is called "rabbit hormone." Selective control of sympathetic and adrenal catecholamine secretion is provided via distinct alpha-2-adrenoceptor subtypes of [25]. In the adrenal medulla of mice, all three alpha(2)-adrenoceptor subtypes (alpha(2A), alpha(2B), and alpha(2C)) play an equal role in the inhibition of noradrenaline overflow, whereas the alpha(2C)-adrenoceptor is the predominant alpha(2)-adrenoceptor subtype involved in the inhibitory mechanism controlling adrenaline overflow [128]. It should also be noted that plasma adrenaline mainly originates from adrenaline-containing cells in the adrenal medulla, whereas plasma noradrenaline reflects not only the release from sympathetic nerves but also the secretion from noradrenaline-containing cells in the adrenal medulla [173].

As noted earlier, CRH plays a central role in the regulation of HPA axis. The action of CRH on ACTH release is strongly potentiated by vasopressin [186]. A1/A2 noradrenergic neurons in the medulla oblongata are well known to mediate stress signals in the central nervous system. Stress activates A1/A2 noradrenergic neurons, and then NA stimulates ACTH secretion through hypothalamic CRH [117]. Serotonin affects the secretion of CRH and ACTH both at the hypothalamic, pituitary portal and pituitary gland levels, and possibly also at the adrenal level [90]. On the other hand, prolactin-releasing peptide was recently isolated and was found to be produced by some A1/A2 neurons and the dorsomedial hypothalamic nucleus. It was shown that prolactin-releasing peptide and NA cooperatively modulate the HPA axis [117].

The results obtained by Stroth N. and Eiden LE. (2010) suggest that sustained corticosterone secretion, synthesis of CRH in the hypothalamus, and synthesis of the enzymes producing the hormone adrenaline in the adrenal medulla, are controlled by pituitary adenylate cyclase-activating polypeptide (PACAP) [181]. PACAP is an important regulator of both central and/or peripheral components of the stress axis and it is thought to act centrally on the paraventricular nucleus of the hypothalamus to regulate both the HPA axis and the sympathetic nervous system. Intriguingly,

14

PACAP is also active in brain structures that mediate anxiety- and fear-related behaviors [77]. PACAP appears to function as an "emergency response" cotransmitter in the sympathoadrenal axis, where the primary secretory response is controlled by a classical neurotransmitter but sustained under paraphysiological conditions by a neuropeptide [73].

According to the literature, androgens influence the growth and differentiation of thyrocytes, increasing expression of androgen receptors, IGF-1 in both males and females, whereas estrogens provide a gender difference which can be realized without expression of estrogen receptors [177]. Sex hormones also modulate the function of the thyroid gland by altering the clearance of thyroxine-binding globulin [187].

Triiodothyronine is the biologically active form of thyroid hormones [140]. About $^2/_3$ circulating T_3 is formed by T_4 deiodination in target tissues, including the brain [94].

Deiodination of T_4 in central nervous tissue is catalyzed by type 2 5'-deiodinase, which normally provides the formation 80% of all T_3 [164]. It is believed that the decrease of type-2 5'-deiodinase activity in depression leads to reducing formation of T_3, which has a powerful antidepressant effect [214]. This hypothesis is confirmed by data about the increase of T_4 and rT_3 levels and reducing T_3 content in cerebrospinal fluid in depression [92].

Thyroid hormones and neurotransmitters on the level of central nervous system form a single neurohumoral system that participates in providing integrative functions of the brain [164]. During fetal development and early neonatal period thyroid hormones stimulate proliferation, differentiation, migration of neurons and glial cells [162], influence processes of synaptogenesis and myelination of nerve fibers [125].

Thyroid hormones have particular importance in the so-called critical period of brain formation, which includes the last trimester of pregnancy and the first weeks after birth [168]. In the brain of adult the thyroid hormones affect the expression of a small number of neuron-specific genes [6], but the main effect of thyroid hormones in the central nervous system of adults is associated with their effects on neurotransmitter neuronal transmission. In particular, thyroid hormones influence serotonin, noradrenaline, dopamine systems [12, 67, 80, 103, 191].

Experimental hypothyroidism in rats is accompanied by the decrease of serotonin concentration in the cerebral cortex of the brain and mezodiencephalon and by the increase of its circuits in the hyppocampus [12, 19, 89]. Administration of thyroid hormones to rats with hypothyroidism and to euthyroid animals leads to an increase in the concentration of serotonin in the cortex [67].

Thyroid hormones also cause desensitization of autoregulatory 5-HT 1A - receptors in neurons of raphe nucleus [67, 129, 191] and increase the density of 5-HT2 receptors in frontal cortex neurons of rats [103, 191]. In turn, serotonin controls the function of the hypothalamic-pituitary-thyroid system at the level of the hypothalamus, inhibiting the synthesis of thyroid hormones [27, 175]. Catecholaminergic axons innervate hypophysiotropic thyrotropin-releasing hormone-synthesizing neurons [58].

Thus, the analysis of literature data suggests the involvement of testosterone and estradiol to the aggression formation, the damage of monoaminergic system neurotransmission in depression and aggression, changing reactivity of HPA axis in these states.

Section 2
MATERIALS AND METHODS

Animals. The work was carried out on 76 male Wistar rats, aging (3, 6 and 12 months), which were held in standard vivarium conditions.

Research was carried out in accordance with the provisions of the International Convention for the Protection of animals used in experiments (Strasbourg, 1985), and the provisions of the Committee on Bioethics of Kharkiv National Medical University, Ministry of Health of Ukraine.

According to the classification of age groups of laboratory animals [217], the 3-month-old rats belong to the second period (puberty, juvenile period), and 6- and 12-month old rats belong to the third, reproductive, period (the young and mature, respectively).

It is known in men two forms of aggression have been described: impulsive and controlled (in animals - spontaneous and adaptive, respectively) [138]. Impulsive aggression is observed in patient with depression [10]. Submissive rats can serve as experimental model of depression. Dominant rats demonstrate adaptive aggression. Distribution of the animals into groups with alternative types of behavior was made using a model of emotional stress "Sensory contact" with some modifications [99, 201]. According to this model, rats were in the individual maintenance for 5 days to prevent the effect of group interaction. After that, they were kept for 2 days in the experimental cages separated by half with perforated transparent partitions, which provided conditions for sensory contact. Testing of the behavior type started 2 days after the animals adapt to new conditions and sensory contact. At the time of testing, partition was removed for 10 minutes.

Testing was carried out for 10 days in the afternoon from 14.00 to 16.00. According to the results of testing, the animals were divided into 3 groups: dominant, balanced and submissive. Dominant males can serve as model of adaptive aggression and submissive ones – as model of spontaneous aggression. In 20 h after the last test, animals were decapitated.

Sample collection. At the time of decapitation, trunk blood was collected into prechilled tubes containing 0.5 M EDTA. Plasma was obtained by centrifugation of blood at 1500 g for 20 min at room temperature. Brains were promptly removed. Brain structures were isolated on ice and frozen at -20°C.

Analysis of hormones. Testosterone, 17β-estradiol, corticosterone, ACTH, triiodothyronine (T_3), thyroxine (T_4) levels were determined in blood plasma by method of immune-enzymatic analysis using "Alkor Bio" (Russia), "DRG ELISAS" and "Assay Designs".

Analysis of monoamines. Contents of serotonin, noradrenaline, dopamine in the rat frontal cortex, hippocampus and cerebellum were determined by fluorometric micromethod [166]. For extraction of monoamines, 50 mg of brain tissue was homogenized in 1.0 ml of HCl-butanol. After centrifugation (10 min at 2000 g), the supernatant was transferred to a tube containing 2.0 ml of heptane and 0.25 ml of 0.1 M HCl. The tube was shaken within 10 minutes, and centrifuged under the same conditions. After phase separation, the aqueous phase was used for determination of biogenic amine contents by fluorometric method on the spectrum fluorimeter MPF-4A, "Hitachi" (Japan). Serotonin content was determined by intrinsic fluorescence. Excitation wavelength was 303 nm, and that of luminescence was 330 nm. After oxidation of cathecholamines the fluorescence was detected (excitation and luminescence wavelengths were respectively 395 and 485 nm to noradrenaline, 330 and 375 nm to dopamine).

Statistical analysis. Statistical analysis of the results was carried out by methods of nonparametric statistics using the package "Statistica 6.0". Nonparametric analogues of dispersion analysis – Kruskal-Wallis and median tests – were used to reveal the dependence of parameters on group. Manna-Whitney test and correlation analysis according to Spearman were used to compare groups in pairs. For all statistical assessments a value of $P < 0.05$ was accepted to be statistically significant.

Section 3 RESULTS

3.1 Gonadal axis: testosterone and 17β-estradiol levels in blood plasma of males with dovinant and submissive types of behavior

Taking into account that testosterone synthesized in the fetal and neonatal testis initiates the process of brain masculinization [165] through its conversion in brain into 17β-estradiol [219] the levels of of testosterone and 17β-estradiol and their ratio in rat males with dominant and submissive types of behavior were investigated. But it should be noted that 17β-estradiol which is produced by testes inhibits the synthesis of androgens either by autocrine or paracrine ways [189]. It is believed that relative but no absolute levels of sex hormones play more important role in etiology of depression [78].

Analysis of blood plasma testosterone level in rats of different age groups but the same behavior type revealed the following predictability: in rat males independently on a behavior type the testosterone level was significantly increased in six-month-old rats versus three-month-old ones and decreased in twelve-month-old rats versus six-month-old ones. No diferences were found between testosterone levels in three- and twelve-month-old rats with the same type of behavior (Table 1).

Table 1

Testosterone level (nM) in rat blood plasma of different age groups with alternative types of behaviour

Age of animals	Median Me	Quartile 25%; 75%
Submissive males		
Three-month-old	3,00	2,50; 4,40
Six-month-old	6,30*	3,90; 10,05
Twelve-month-old	2,70**	1,55; 4,05
Balanced males		
Three-month-old	7,30	6,60; 9,80
Six-month-old	17,80*	17,40; 21,30
Twelve-month-old	7,90**	7,30; 8,30
Dominant males		
Three-month-old	20,90	19,70; 30,60
Six-month-old	44,90*	38,70; 48,00
Twelve-month-old	16,05**	14,20; 19,400

*- P<0.05 versus three-month-old males; **-P<0.05 versus six-month-old males.

In all investigated age groups the testosterone level was revealed to depend on behavior type. It was increased in dominant rat males versus both balanced and submissive ones. Testosterone level in balanced rats was higher than in submissive males but lower than in dominant ones. The most considerable differences were observed in six-month-old rats (Fig. 1-3).

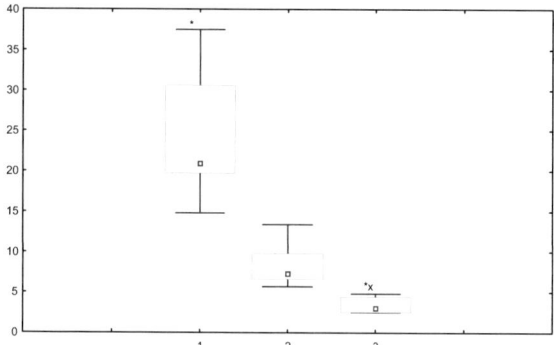

Figure 1 Testosterone levels (nM) in blood plasma of three-month-old rat males with dominant (1), balanced (2) and submissive (3) types of behavior (Me [25%; 75%], min and max). *-P<0.05 versus the rats with balanced type of behavior; x-P<0.05 versus the rats with dominant type of behavior.

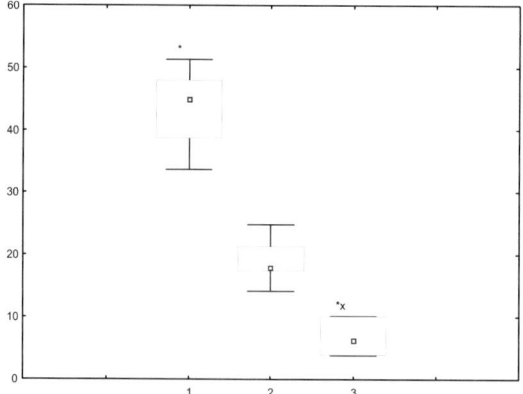

Figure 2 Testosterone levels (nM) in blood plasma of six-month-old rat males with dominant (1), balanced (2) and submissive (3) types of behavior (Me [25%; 75%], min and max). *-P<0.05 versus the rats with balanced type of behavior; x-P<0.05 versus the rats with dominant type of behavior.

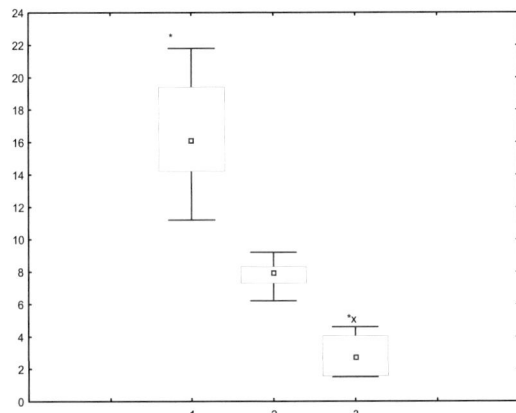

Figure 3 Testosterone levels (nM) in blood plasma of twelve-month-old rat males with dominant (1), balanced (2) and submissive (3) types of behavior (Me [25%; 75%], min and max). *-P<0.05 versus the rats with balanced type of behavior; x-P<0.05 versus the rats with dominant type of behavior.

In contrast to testosterone the level of blood plasma 17β-estradiol was revealed to decrease in dominant rat males versus both submissive and balanced ones (Fig. 4-6). The level of 17β-estradiol in balanced rats was higher than in dominant males but lower than in submissive ones.

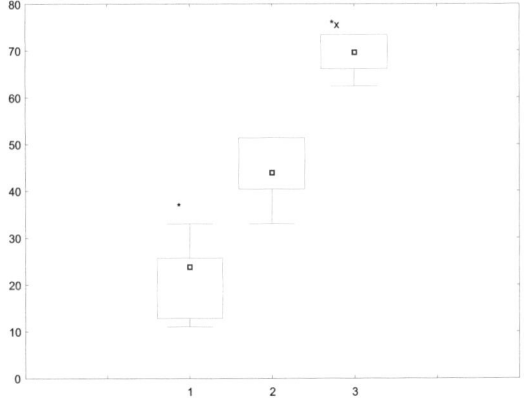

Figure 4 17β-estradiol levels (pM) in blood plasma of three-month-old rat males with dominant (1), balanced (2) and submissive (3) types of behavior (Me [25%; 75%], min and max). *-P<0.05 versus the rats with balanced type of behavior; x-P<0.05 versus the rats with dominant type of behavior.

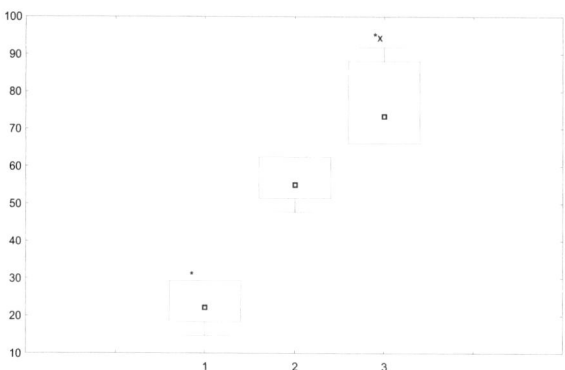

Figure 5 17β-estradiol levels (pM) in blood plasma of six-month-old rat males with dominant (1), balanced (2) and submissive (3) types of behavior (Me [25%; 75%], min and max). *-P<0.05 versus the rats with balanced type of behavior; x-P<0.05 versus the rats with dominant type of behavior.

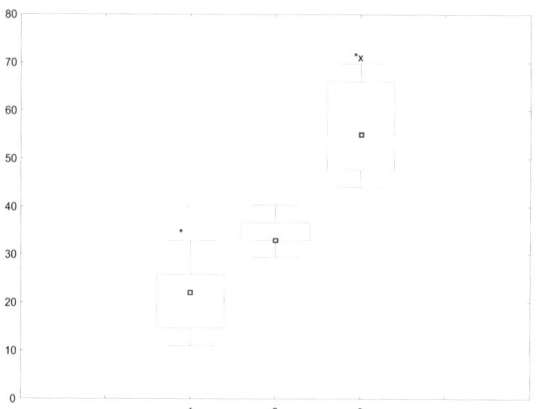

Figure 6 17β-estradiol levels (pM) in blood plasma of twelve-month-old rat males with dominant (1), balanced (2) and submissive (3) types of behavior (Me [25%; 75%], min and max). *-P<0.05 versus the rats with balanced type of behavior; x-P<0.05 versus the rats with dominant type of behavior.

Statistically significant negative correlation between testosterone and 17β-estradiol levels was revealed in all investigated age groups regardless of the type of behavior (Table 2).

Table 2

Correlation between testosterone and 17β-estradiol levels in blood plasma of rat males with different types of behavior

Age Type of behavior	Three-month-old	Six-month-old	Twelve-month-old
Balanced	-0.95*	-0.88*	-0.69*
Submissive	-0.98*	-0.92*	-0.96*
Dominant	-0.92*	-0.92*	-0.99*

* - P<0.05 Correlation is statistically significant.

Statistically significant differences of testosterone/17β-estradiol ratio were found between males with alternative types of behavior (Table 3). The highest ratio of testosterone to 17β-estradiol was found in dominant rat males.

Table 3

Testosterone/17β-estradiol ratio in rat males with different types of behavior

Type of behaviour	Three-month-old	Six-month-old	Twelve-month-old
Submissive	43.01	41.40	49.03
Dominant	844.11	2008.05	759.07
Balanced	165.75	346.37	233.05

3.2 Hypothalamic pituitary adrenal axis: adrenocorticotropic and corticosterone levels in blood plasma of males with dominant and submissive types of behavoir

Androgens play a crucial role in regulation of HPA axis [111]. In this regard we have investigated the correlation between the content of testosterone and levels of HPA axis components (adrenocorticotropic hormone – ACTH and corticosterone) in rats with dominant and submissive types of behaviour.

Blood plasma ACTH levels in rats of different age groups but the same behavior type were not changed. The dependence of blood plasma ACTH level in rats of investigated groups on type of behavior was revealed. In all age groups of submissive animals ACTH level in blood plasma was higher, compared with balanced and dominant ones. In all age groups the difference between ACTH levels in balanced and dominant males was not found (Fig. 7).

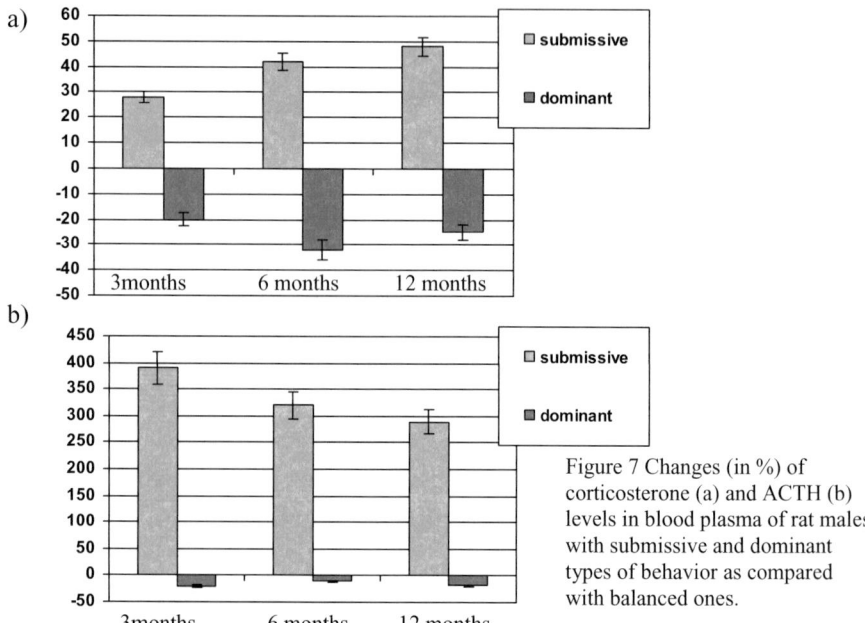

Figure 7 Changes (in %) of corticosterone (a) and ACTH (b) levels in blood plasma of rat males with submissive and dominant types of behavior as compared with balanced ones.

Corticosterone level in blood plasma of males of different age groups but the same behavior type was not changed. The dependence of blood plasma corticosterone level on behavior type was revealed in all investigated age groups. In all age groups of submissive animals the blood plasma corticosterone level was higher, compared with balanced and dominant ones, but in balanced males it was higher than in dominant males.

Strong positive correlations between ACTH and corticosterone levels were observed in all age groups of submissive rats, in balanced animals of the young and mature reproductive period, but in dominant rats the correlation between these parameters was statistically insignificant.

Strong negative correlations between testosterone and ACTH (or corticosterone) levels were found in all investigated groups, but in dominant rats of young reproductive period the correlation between testosterone and corticosterone levels was statistically insignificant (Fig. 8).

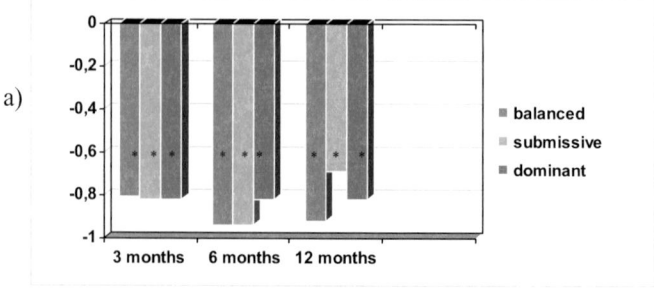

b)

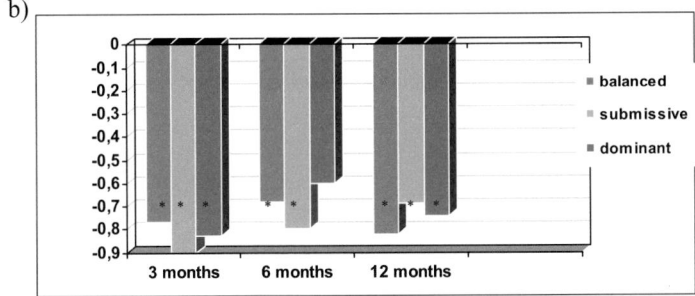

Figure 8 Correlation coefficients between testosterone and ACTH (a) or corticosterone (b) levels in blood plasma of rat males of different age groups with alternative behavior types

Taken together, these results indicate HPA axis hyperactivity in submissive males and suggest the influence of testosterone on the formation and functioning of HPA axis.

3.3 Thyroid axis: triiodothyronine, thyroxine levels in blood plasma of males with dominant and submissive types of behavior

The difference between contents of T_3 in blood plasma was observed only in rats of young reproductive period: in submissive males it was lower than in both balanced and dominant ones (Fig. 9). The difference between the balanced and dominant rats was not revealed.

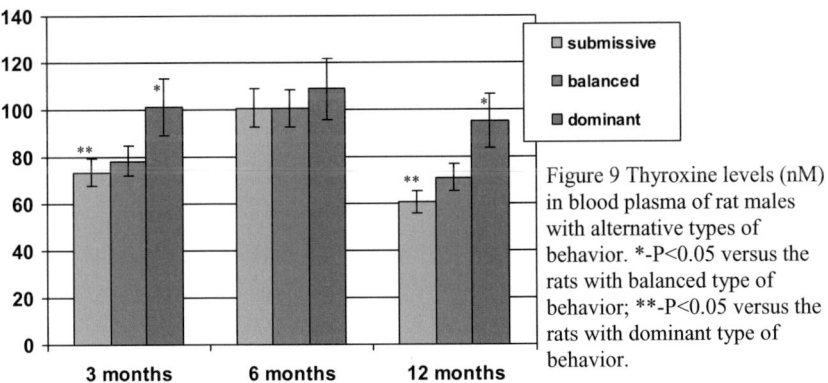

Figure 9 Thyroxine levels (nM) in blood plasma of rat males with alternative types of behavior. *-P<0.05 versus the rats with balanced type of behavior; **-P<0.05 versus the rats with dominant type of behavior.

No difference in blood plasma T_4 level was found in rats of young reproductive period. In other age groups the level of thyroxine was higher in dominant animals compared to submissive ones (Fig. 10).

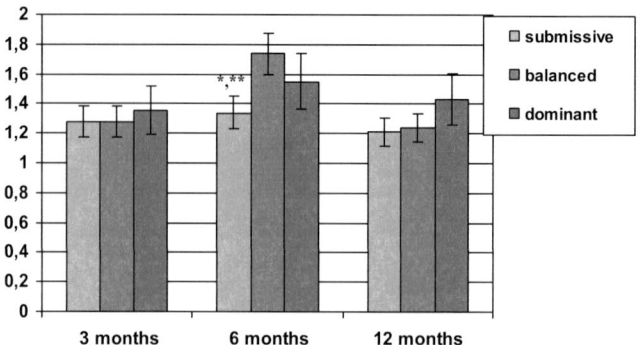

Figure 10 Triiodothyronine levels (nM) in blood plasma of rat males with alternative types of behavior. *-P<0.05 versus the rats with balanced type of behavior; **-P<0.05 versus the rats with dominant type of behavior.

Correlation analysis of the investigated parameters in dominant rats showed high positive correlation between testosterone and thyroxine in all age groups, between testosterone and T_3 in all age groups, but in the juvenile period a correlation was not statistically significant (Tables 4, 5).

Table 4

Correlation between testosterone and thyroxine levels in blood plasma of rat males with different types of behavior

Type of behaviour	Three-month-old	Six-month-old	Twelve-month-old
Balanced	-0.46	+0.65	-0.33
Submissive	+0.71	+0.89*	+0.50
Dominant	+0.91*	+0.80*	+0.81*

* - P<0.05 Correlation is statistically significant.

In submissive rats no statistically significant correlation was revealed between testosterone and T_4 and between testosterone and T_3, except for young reproductive period males. In submissive males of young reproductive period a strong positive correlation between the levels of testosterone and thyroxine (Table 4) and a strong negative correlation between the levels of testosterone and triiodothyronine were observed (Table 5).

In balanced animals no correlation between the studied parameters was found (Table 4, 5).

Table 5

Correlation between testosterone and triiodothyronine levels in blood plasma of rat males with different types of behavior

Type of behavior	Three-month-old	Six-month-old	Twelve-month-old
Balanced	-0.28	-0.57	-0.23
Submissive	+0.44	-0.85*	+0.17
Dominant	+0.69	+0.78*	+0.73*

* - $P<0.05$ Correlation is statistically significant.

Correlation between the contents of hormones T_3 and T_4 was absent in males of all age groups and different types of behavior, with exception of dominant rats of mature reproductive age ($r=0.76$, $P<0.05$).

In balanced rats no statistically significant correlations between components of HPA axis and thyroid hormones were revealed, with exception of a positive correlation between corticosterone and triiodothyronine ($r=+0.68$, $P<0.05$).

In submissive rats statistically significant correlations between components of HPA axis and thyroid hormones were observed only in young reproductive age: close positive correlation between T_3 and ACTH (or corticosterone) ($r=+0.93$ and $r=+0.94$, respectively, $P<0.05$); negative correlation between T_4 and ACTH ($r=-0.75$, $P<0.05$); between T_4 and corticosterone correlation ($r=-0.63$) was not statistically significant.

In dominant males the correlations between the levels of T_3 and HPA axis components were not found; strong statistically significant correlation between T_4 and ACTH was observed in juvenile and young reproductive periods ($r=-0.77$ and $r=-0.82$ respectively, $P<0.05$; high negative correlation between T_4 and corticosterone was identified in juvenile and mature reproductive periods ($r=-0.85$ and $r=-0.93$ respectively, $P<0.05$).

Thus, not all age groups revealed changes of thyroid hormones depending on the behavior. If these changes were observed, the level of thyroid hormones was higher in dominant animals. Namely in dominant animals a statistically significant positive correlation between testosterone and thyroid hormones was observed. Thyroxine is the main hormone that is synthesized by thyroid gland [140]. According to our data, the dominant rats revealed a high positive correlation between testosterone and thyroxine, which indicates the possibility of a positive influence of testosterone on the synthesis of thyroid hormones in the male with dominant type of behavior. In dominant rats, compared with rats other types of behavior, higher correlations between T_4 and ACTH levels were observed.

3.4 Monoaminergic system: serotonin, noradrenaline and dopamine contents in frontal cortex, hippocampus and cerebellum of males with dominant and submissive types of behavior

Monoaminergic brain systems have a great influence on the emotions and behavior in humans and animals [27, 132, 198]. Neocortex and the hippocampus play a crucial role in the development of the symptoms relating to cognitive deficiency, which is manifested in depressive patients [47]. The frontal cortex is also involved in the control of anxiety, emotions and behavior [53]. It is one of the structures related to the control of aggressive behavior [4, 148]. In this regard, we have investigated serotonin, noradrenaline, dopamine contents in the frontal cortex and the hippocampus in different age groups of rats with dominant and submissive types of behavior [146].

According to the results obtained, in the frontal cortex and the hippocampus of submissive males of all ages, serotonin content was significantly lower, compared to that in the balanced and dominant rats (Table 6).

Table 6

Serotonin content (nmol/g of tissue) in brain of rats with alternative behavior types

Behavior type	Three-month-old		Six-month-old		Twelve-month-old	
	Median Me	Quartile 25%; 75%	Median Me	Quartile 25%; 75%	Median Me	Quartile 25%; 75%
Frontal cortex						
Submissive	2.38*,**	2.37; 5.97	7.26*,**	5.59; 9.85	2.98*,**	2.98; 7.46
Balanced	14.93	10.75; 17.92	16.72	14.33; 20.90	13.14	13.14; 16.72
Dominant	14.63	14.33; 20.90	18.51	16.72; 20.90	18.22*	17.32; 18.81
Hippocampus						
Submissive	5.15*,**	4.12; 7.73	8.51*,**	7.51; 11.09	4.12*,**	3.61; 7.99
Balanced	13.41	11.86; 16.50	17.54	14.96; 21.15	14.96	14.44; 18.57
Dominant	16.25*	15.99; 21.66	19.08	18.57; 21.66	19.24*	18.31; 21.15

*-$P < 0.05$ versus the rats with balanced type of behavior; **$P < 0.05$ versus the rats with dominant type of behavior

28

No differences of serotonin content between animals with a dominant and a balanced type of behavior in young reproductive period were found. The serotonin content difference between balanced and dominant males was observed in juvenile (hippocampus) and mature reproductive (frontal cortex and hippocampus) period (Table 6).

A strong negative correlation between the content of testosterone in the blood plasma and serotonin in the brain (hippocampus and frontal cortex) in two age groups of balanced males (in the juvenile and young reproductive period) was revealed. In balanced males of mature reproductive period, correlation was not observed (Table7).

Table 7

Correlation coefficients between testosterone level in blood plasma and serotonin content in brain regions of different age group rats with alternative behavior types. *P < 0.05

Group of animals	Three-month-old	Six-month-old	Twelve-month-old
Frontal cortex			
Balanced	-0.75*	-0.75*	-0.22
Submissive	-0.48	+ 0.74*	+ 0.92*
Dominant	+ 0.64*	+ 0.79*	+ 0.76*
Hippocampus			
Balanced	-0.71*	-0.69*	-0.15
Submissive	-0.43	+ 0.74*	+ 0.88*
Dominant	+ 0.48	+ 0.76*	+ 0.73*

In rats of the young and mature reproductive period of dominant and submissive types of behavior, there was a close positive correlation between the content of testosterone in the blood plasma and serotonin in the frontal cortex and the hippocampus (Table 7).

Correlation between the content of serotonin in the frontal cortex (or hippocampus) and HPA system components was observed mainly in young reproductive age. Positive correlations in balanced rats were changed to negative ones in submisive and dominant males, and in submissive animal of this period correlation was statistically significant (Table 8).

In submissive rats of young and mature reproductive periods a close positive correlation between T_4 level in blood plasma and serotonin content in the frontal cortex as well as hippocampus was found. In balanced animals of young reproductive period a strong negative correlation between T_4 in blood plasma and serotonin content both in the frontal cortex and hippocampus was revealed (Table 9). In dominant rats statistically significant correlation between the level of T_4 in the blood plasma and serotonin content in the frontal cortex in juvenile period, between the level of T_4 in the blood plasma and serotonin content in the hippocampus in mature

Table 8

Correlation coefficients between corticosterone (or ACTH) level in blood plasma and serotonin content in brain regions of different age group rats with alternative behavior types. *P < 0.05

Group of animals	Frontal cortex			Hippocampus		
	Corticosterone - serotonin					
	3 months	6 months	12 months	3 months	6 months	12 months
Balanced	+0.62	+0.65	+0.30	+0.60	+0.60	+0.24
Submissive	+0.73	-0.72*	-0.36	+0.46	-0.80*	-0.24
Dominant	-0.79*	-0.46	-0.33	-0.58	-0.59	-0.57
	ACTH - serotonin					
Balanced	+0.53	+0.74*	+0.47	+0.53	+0.76*	-0.03
Submissive	+0.67	-0.82*	-0.44	+0.45	-0.84*	-0.29
Dominant	-0.34	-0.66	-0.60	-0.27	-0.61	-0.52

reproductive age. Strong positive correlation between the level of T_3 in blood plasma and serotonin content in the frontal cortex in dominant rats of juvenile and young reproductive periods and between T_3 level in blood plasma and serotonin content in

Table 9

Correlation coefficients between thyroxine level in blood plasma and serotonin content in brain regions of different age group rats with alternative behavior types. *P < 0.05

Group of animals	Three-month-old	Six-month-old	Twelve-month-old
	Frontal cortex		
Balanced	+0.10	-0.85*	-0.01
Submissive	+0.15	+0.98*	+0.76*
Dominant	+0.74*	+0.50	+0.47
	Hippocampus		
Balanced	+0.28	-0.72*	-0.15
Submissive	+0.37	+0.96*	+0.86*
Dominant	+0.67	+0.50	+0.75*

the hippocampus of dominant animals of young reproductive period was revealed (Table 10). In submissive rats a correlation between level of T_3 and serotonin content in the frontal cortex (or hippocampus) was observed only in young reproductive period; it was negative.

Table 10

Correlation coefficients between triiodothyronine level in blood plasma and serotonin content in brain regions of different age group rats with alternative behavior types. *P < 0.05

Group of animals	Three-month-old	Six-month-old	Twelve-month-old
Frontal cortex			
Balanced	+0.18	+0.61	+0.21
Submissive	-0.31	-0.75*	-0.03
Dominant	+0.75*	+0.87*	+0.55
Hippocampus			
Balanced	+0.18	+0.70*	+0.15
Submissive	-0.04	-0.83*	-0.16
Dominant	+0.36	+0.87*	+0.66

In balanced rats of young reproductive period a statistically significant positive correlation between the level of T_3 and serotonin content in the hippocampus was revealed.

In all age groups the content of norepinephrine in both the hippocampus and frontal cortex was higher in submissive animals compared to the balanced and dominant ones (Table 11).

Table 11

Noradrenaline content (nmol/g of tissue) in brain of rats with alternative behavior types

Behavior type	Three-month-old		Six-month-old		Twelve-month-old	
	Median Me	Quartile 25%; 75%	Median Me	Quartile 25%; 75%	Median Me	Quartile 25%; 75%
Frontal cortex						
Submissive	7.36*,**	6.97; 8.51	7.88*,**	5.81; 9.21	5.42*,**	4.65; 7.56
Balanced	3.87	2.71; 4.26	4.65	4.65; 5.03	3.87	3.48; 4.26
Dominant	2.32*	2.13; 3.10	2.32*	2.32; 3.48	3.10*	2.32; 3.29
Hippocampus						
Submissive	9.16*,**	7.23; 10.13	8.92*,**	6.27; 9.48	9.16*,**	7.96; 9.65
Balanced	4.82	4.34; 6.24	5.79	4.34; 5.79	5.30	4.82; 5.79
Dominant	4.34	3.89; 5.31	3.37*	2.89; 4.34	4.10*	3.62; 4.34

*P < 0.05 versus the rats with balanced type of behavior; **P < 0.05 versus the rats with dominant type of behavior

In all age groups, with the exception of juvenile period males (hippocampus), there was a statistically significant difference between the content of norepinephrine in the dominant and balanced animals (Table 11). In the balanced males noradrenaline content was higher compared to the dominant ones.

There was no correlation between the level of testosterone in blood plasma and content of norepinephrine in the frontal cortex in all groups studied, except submissive and dominant rats of young reproductive period in which a close negative correlation between these parameters was found (Table 12).

In the hippocampus, the structure that regulates the duration of response to stress, there was a strong negative correlation between these parameters in all groups of animals, with the exception of dominant rats of juvenile period and balanced and submissive rats of mature reproductive period (Table 12), that may be one of the mechanisms to reduce by androgens a sensitivity to stress.

Table 12

Correlation coefficients between testosterone level in blood plasma and noradrenaline content in brain regions of different age group rats with alternative behavior types. *P < 0.05

Group of animals	Three-month-old	Six-month-old	Twelve-month-old
Frontal cortex			
Balanced	+0.09	-0.46	-0.60
Submissive	-0.17	-0.67*	-0.54
Dominant	-0.45	-0.64*	-0.52
Hippocampus			
Balanced	-0.69*	-0.72*	-0.55
Submissive	-0.79*	-0.69*	-0.52
Dominant	-0.61	-0.73*	-0.72*

Attention is drawn to the fact that the correlation coefficients between the HPA axis components and noradrenaline in all the cases, except for the coefficient of correlation between the level of plasma ACTH and noradrenaline content in the frontal cortex of juvenile period balanced rats, have a positive sign. However, in the frontal cortex the statistically significant positive correlations were revealed only in the dominant and submissive males of young reproductive period between the contents of ACTH and norepinephrine and in the dominant males of mature reproductive period between corticosterone and noradrenaline contents (Table 13). A close positive correlation between the level of plasma ACTH and norepinephrine content in the hippocampus was observed in all groups except the males of mature reproductive period and dominant males of juvenile period. The correlation between the corticosterone level in the plasma and norepinephrine content in the hippocampus was observed in balanced animals of mature reproductive period and submissive and dominant rats of juvenile period (Table 9). This correlation was also positive.

Table 13

Correlation coefficients between corticosterone (or ACTH) level in blood plasma and noradrenaline content in brain regions of different age group rats with alternative behavior types. *P < 0.05

Group of animals	Frontal cortex			Hippocampus		
	Corticosterone - noradrenaline					
	3 months	6 months	12 months	3 months	6 months	12 months
Balanced	+0.38	+0.48	+0.63	+0.53	+0.65	+0.73*
Submissive	+0.11	+0.57	+0.53	+0.88*	+0.60	+0.58
Dominant	+0.24	+0.21	+0.86*	+0.79*	+0.32	+0.55
	ACTH - noradrenaline					
Balanced	-0.15	+0.55	+0.47	+0.89*	+0.77*	+0.53
Submissive	+0.01	+0.73*	+0.22	+0.88*	+0.79*	+0.20
Dominant	+0.59	+0.88*	+0.39	+0.27	+0.90*	+0.65

Calculating the coefficient of correlation between the level of T_3 in blood plasma and norepinephrine content in the frontal cortex (or hippocampus) demonstrated the presence of a positive correlation between these parameters in balanced males of mature reproductive age (r=+0.70 and r=+0.81 respectively, P<0.05). Correlation between the level of T_4 in the blood plasma and norepinephrine content in the frontal cortex was found only in balanced males of mature reproductive age (strong positive correlation: r=+0.79, P<0.05), and dominant rats of young and mature reproductive periods (strong negative correlation: r=-0.69 and r=-0.76 respectively, P<0.05). In the hippocampus a positive correlation between the level of T_4 in blood plasma and norepinephrine content was observed in balanced rats of the juvenile period (r=+0.80, P<0.05).

The content of dopamine in submissive males was lower compared with both dominant and balanced rats. The difference between the content of dopamine in the frontal cortex in the balanced and dominant males was observed only in young reproductive period, and in the hippocampus it was revealed only in mature reproductive age (Table 14).

Table 14

Dopamine content (nmol/g of tissue) in brain of rats with alternative behavior types

Behavior type	Three-month-old		Six-month-old		Twelve-month-old	
	Median Me	Quartile 25%; 75%	Median Me	Quartile 25%; 75%	Median Me	Quartile 25%; 75%
Frontal cortex						
Submissive	6.52*,**	3.26; 9.79	11.42*, **	9.79 14.68	7.34*, **	4.89; 9.79
Balanced	37.53	32.63; 45.69	32.63	32.63; 42.43	39.16	22.84; 40.79
Dominant	26.11	20.40; 37.53	26.11*	22.84; 29.37	36.71	27.74; 44.79
Hippocampus						
Submissive	9.32*, **	7.46; 14.92	13.99*, **	11.19; 16.78	10.26*, **	8.39; 12.12
Balanced	41.03	37.20; 48.49	37.30	37.30; 48.49	41.03	27.97; 44.46
Dominant	52.22	42.90; 58.50	48.49	44.76; 48.49	54.08*	48.49; 61.05

*P < 0.05 versus the rats with balanced type of behavior; **P < 0.05 versus the rats with dominant type of behavior

In the balanced animals of juvenile and young reproductive periods the correlations between blood plasma testosterone level and dopamine content in the hippocampus (or frontal cortex) were positive, but only in young reproductive age the correlation was statistically significant (Table 15).

Table 15

Correlation coefficients between testosterone level in blood plasma and dopamine content in brain regions of different age group rats with alternative behavior types. *P < 0.05

Group of animals	Three-month-old	Six-month-old	Twelve-month-old
Frontal cortex			
Balanced	+0.58	+0.89*	-0.48
Submissive	-0.49	-0.71*	-0.04
Dominant	+0.86*	+0.51	+0.57
Hippocampus			
Balanced	+0.52	+0.89*	-0.49
Submissive	-0.49	-0.89*	-0.16
Dominant	+0.69 P=0.058	+0.64	+0.64

In the dominant males of all investigated age groups the correlations between these parameters were also positive, but they were statictically significant only in juvenil period (frontal cortex). It should be noted that in submissive rats these correlations were negative and statistically significant in young reproductive period (Table 15).

Analysis of correlations between HPA axis component levels in blood plasma and dopamine contents in both the hippocampus and frontal cortex revealed the following pattern: in young reproductive period in the balanced males the correlation was negative, in the submissive males it was positive, in the dominant males it was not found (Table 16).

In the juvenile period, statistically significant correlation was observed only between ACTH level in blood plasma and dopamine content in the frontal cortex of the dominant rats. This correlation was negative. In mature reproductive period positive correlation between blood plasma corticosterone level and dopamine content (both in the frontal cortex and hippocampus) in the balanced males and negative correlation between blood plasma corticosterone level and dopamine content in hippocampus in the dominant males were observed (Table 16).

Table 16

Correlation coefficients between corticosterone (or ACTH) level in blood plasma and dopamine content in brain regions of different age group rats with alternative behavior types. *P < 0.05

Group of animals	Frontal cortex			Hippocampus		
	Corticosterone - dopamine					
	3 months	6 months	12 months	3 months	6 months	12 months
Balanced	-0.43	-0.83*	+0.78*	-0.52	-0.83*	+0.78*
Submissive	+0.57	+0.73*	+0.08	+0.57	+0.82*	+0.08
Dominant	-0.57	-0.24	-0.69 P=0.058	-0.33	-0.28	-0.86*
	ACTH - dopamine					
Balanced	-0.33	-0.83*	+0.50	-0.58	-0.83*	+0.50
Submissive	+0.45	+0.79*	-0.34	+0.45	+0.96*	-0.26
Dominant	-0.96*	-0.24	-0.29	-0.68	-0.34	-0.37

A strong correlation between T_3 blood plasma level and dopamine content in frontal cortex (or hippocampus) was observed only in submissive males of young reproductive period (r=+0.83 and r=+0.91 respectively, P<0.05).

The analysis of correlation between the level of T_4 in the blood plasma and content of dopamine in the brain showed the following pattern: in young reproductive period a strong correlation was observed in rats of all types of behavior, but in dominant rats the correlation between the level of T_4 in the blood plasma and dopamine content in the frontal cortex was not statistically significant (Table 17).

The correlation was positive in the balanced and dominant males and was negative in submissive rats. In addition, a strong negative correlation between the level of T_4 in the blood plasma and dopamine content in the hippocampus was also revealed in submissive animals of mature reproductive period. A close positive correlation between T_4 and dopamine (both in the hippocampus and frontal cortex) in the dominant rats of the same age and between the level of T_4 in blood plasma and dopamine content in the frontal cortex in dominant rats of juvenile period was shown (Table 17).

Table 17

Correlation coefficients between thyroxine level in blood plasma and dopamine content in brain regions of different age group rats with alternative behavior types. *P < 0.05

Group of animals	Three-month-old	Six-month-old	Twelve-month-old
Frontal cortex			
Balanced	-0.12	+0.72*	+0.11
Submissive	+0.15	-0.81*	-0.67
Dominant	+0.81*	+0.56	+0.76*
Hippocampus			
Balanced	+0.20	+0.72*	+0.07
Submissive	+0.15	-0.79*	-0.82*
Dominant	+0.43	+0.73*	+0.92*

The investigation of monoamine contents in cerebellum of rats with alternative types of behavior revealed the same direction of monoamine content changes in cerebellum as in the regions of emotiogenic limbicocortical system of brain:

• the increase of noradrenaline content in submissive males, the decrease of this parameter in dominant males versus the balanced ones (Fig. 11);

• reducing dopamine content in rats with submissive type of behavior versus balanced and dominant animals (Fig. 12);

• the decrease of serotonin content in submissive rats versus both balanced and dominant males (Fig. 13).

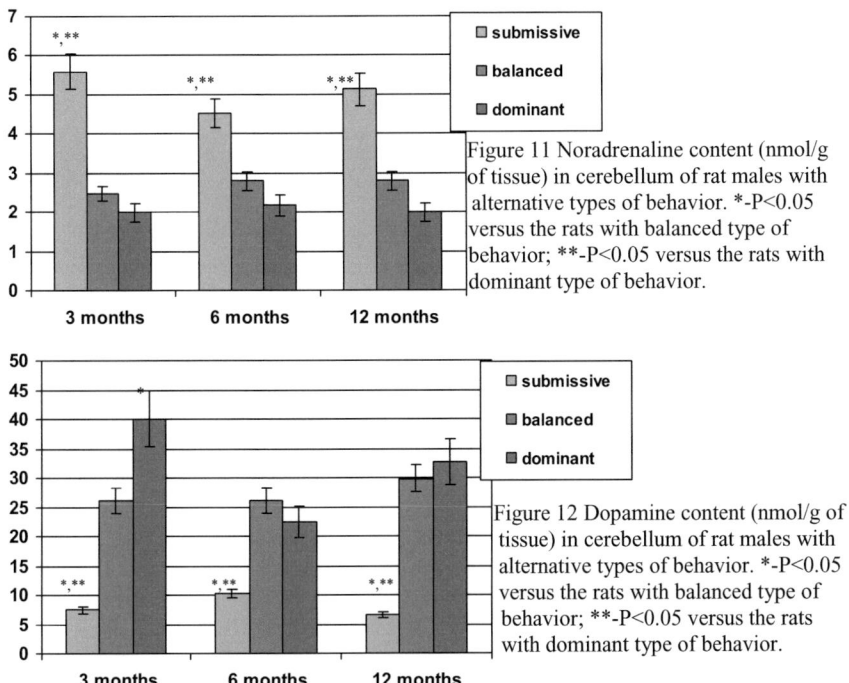

Figure 11 Noradrenaline content (nmol/g of tissue) in cerebellum of rat males with alternative types of behavior. *-P<0.05 versus the rats with balanced type of behavior; **-P<0.05 versus the rats with dominant type of behavior.

Figure 12 Dopamine content (nmol/g of tissue) in cerebellum of rat males with alternative types of behavior. *-P<0.05 versus the rats with balanced type of behavior; **-P<0.05 versus the rats with dominant type of behavior.

Correlation analysis between catecholamine contents in cerebellum and the regions of emotiogenic limbicocortical system revealed a strong positive correlation betveen noradrenaline contents in cerebellum and frontal cortex in dominant males of

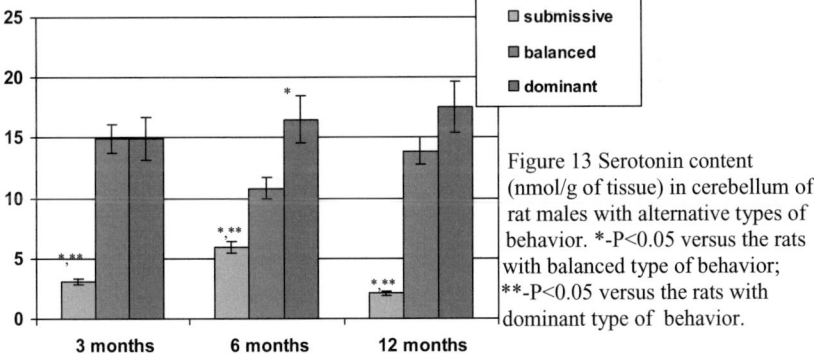

Figure 13 Serotonin content (nmol/g of tissue) in cerebellum of rat males with alternative types of behavior. *-P<0.05 versus the rats with balanced type of behavior; **-P<0.05 versus the rats with dominant type of behavior.

young reproductive period (r = +0.746, P<0.05); between noradrenaline contents in cerebellum and hippocampus in dominant rats of young and mature reproductive periods (r = +0.704 and r = +0.798 respectively, P<0.05) and in balanced rats of mature reproductive period (r = +0.867; P<0.05). Moreover, strong positive correlation between dopamine contents in frontal cortex and cerebellum in both dominant and submissive rats, between dopamine contents in hippocampus and cerebellum of practically all investigated groups was observed (Table 18).

Table 18

Correlation coefficients between dopamine contents in cerebellum and frontal cortex/hippocampus of rats with alternative types of behavior, *-P < 0.05

Type of behavior	Three-month-old	Six-month-old	Twelve-month-old
Cerebellum – Frontal cortex			
Balanced	+0.400	+0.637	+0.613
Submissive	+0.791*	+0.881*	+0.869*
Dominant	+0.792*	+0.779*	+0.351
Cerebellum – Hippocampus			
Balanced	+0.821*	+0.637	+0.667 P=0.051
Submissive	+0.891*	+0.887*	+0.829*
Dominant	+0.690P=0.058	+0.683*	+0.280

Strong positive correlations were revealed between serotonin contents in cerebellum and hippocampus in all investigated groups excepting dominant rats of mature reproductive period. Strong positive correlations were also observed between serotonin contents in cerebellum and frontal cortex of male of juvenil and young reproductive periods of all behavior types and of mature reproductive period males with dalanced type of behavior (Table 19).

Table 19

Correlation coefficients between serotonin contents in cerebellum and frontal cortex/hippocampus of rats with alternative types of behavior, *-P < 0.05

Type of behavior	Three-month-old	Six-month-old	Twelve-month-old
Cerebellum – Frontal cortex			
Balanced	+0.854*	+0.904*	+0.846*
Submissive	+0.820*	+0.720*	+0.664
Dominant	+0.732*	+0.804*	+0.310
Cerebellum – Hippocampus			
Balanced	+0.750*	+0.829*	+0.842*
Submissive	+0.938*	+0.804*	+0.852*
Dominant	+0.846*	+0.733*	+0.542

Section 4   **DISCUSSION**

The data obtained in our studies, namely the revealed correlation between testosterone level and type of behavior, the highest level of testosterone in dominant adulthood males, are associated with the role of testosterone in processes of sexual differentiation and development of brain [161]. Testosterone is the principal male sex steroid which is involved in the regulation of aggression both in animals, and in humans [209]. In sensitive period of early fetal and perinatal development testicular androgens influence neuroendocrine system and mechachisms of brain differentiation providing gender-specific behaviour [85]. During embryonal period the influence of testosterone on hypothalamus predetermines the character of gonadotropin secretion in adults and therefore the specific proportion between male and female sex hormones. In perinatal period androgens facilitate the formation of neuronal networks which are expressed by aggressive behavior in adults [37]. In adulthood testosterone is suggested to contribute to the modulation of neuronal ways which regulate aggression [132].

There are evidences of relationship between circulating testosterone and aggression, for example in pubertic period when testosterone and aggression levels increase parallely in most mammalian species [96]. In adolescent men the aggression correlates with increased level of testosterone in blood plasma [185]. There is positive correlation between testosterone level and criminal behaviour [197].

The conversion of androgens into estrogens in the brain is the key mechanism by which testosterone regulates many physiological and behavioral processes throughout the animal life [161]. Neonatal aromatization of testosterone into 17β-estradiol is the major factor responsible for development of adult aggression between males [64]. In adult males the aggression also depends on circulating testicular hormones [134] and may be mostly, if not entirely, mediated through activation of estrogen receptors (ER), with enhancement of aggression via ERα activation and inhibition via ERβ activation [134, 167].

It should be noted that in the postnatal and adult brain regions the content of 17β-estradiol considerably decreases. However the endogenous content of 17α-estradiol and estrone in the postnatal and adult neocortex, hippocampus, hypothalamus, and cerebellum of both sexes is significantly elevated compared with 17β-estradiol. 17α-Estradiol is not involved in expression of aggression. It has strong neuroprotective activity [59, 106], is able to induce short-latency effects on spatial memory through influences on hippocampal synaptic plasticity [112]. The synthesis of 17α-estradiol in brain does not depend on gonadal steroids. Gonadectomy and adrenalectomy does not influence 17α-estradiol content [192].

According to our data, the level of blood plasma 17β-estradiol was decreased in dominant rat males versus both submissive and balanced ones. The level of 17β-estradiol in balanced rats was higher than in dominant males but lower than in submissive ones.

Very low levels of 17β-estradiol in adult brain regions and adrenals [192] testify to testicular origin of blood plasma 17β-estradiol. This suggestion is confirmed by revealed in our study the negative correlation between testosterone and 17β-estradiol levels since production of estrogens in testis inhibits the synthesis of androgens.

Considerable difference between testosterone/17β-estradiol ratio in males with dominant and submissive types of behavior may be used to determine the predisposition of organism to different types of aggression and to depression.

As androgens play a critical role in the regulation of HPA axis [110], and its dysregulation is often associated with anxiety and depression, as well as violence and abnormal aggression, we have investigated the HPA axis components, such as ACTH and corticosterone levels in rats with alternative types of behavior.

The results indicate the increased activity of HPA axis in submissive rats. In all age groups in submissive animals ACTH and corticosterone levels in blood plasma were higher compared to the balanced and dominant males.

Testosterone can influence the HPA axis both in the embryonic period and adulthood [186]. Testosterone triggers sexual differentiation in the critical period of fetal and perinatal development, converting in the brain to 17β-estradiol [161]. Testosterone can organize normal functioning HPA axis, in which stress induced elevations in ACTH and corticosterone release decline over repeated exposures to the same stimulus [17]. The findings obtained by Bingham B. et al. (2011) confirm an organizing influence of both androgen receptors and androestrogen conversion on HPA axis response to repeated psychogenic stress. In these studies a significant reduction of corticosterone after repeated stress was shown in males that were castrated as adults, but the reduction was not observed in adult male rats that received subcutaneous capsules containing the antiandrogen flutamide or the aromatase inhibitor, introduced within 12h of birth and removed on day 21. All groups demonstrated smaller declines of ACTH in response to repeated stress compared to control animals [17].

The presence of a strong negative correlation between testosterone and components of HPA axis found in our study confirms the important role of androgens in the organizing and regulation of HPA axis and demonstrates the inhibitory effect of testosterone on this system.

One of the most important structures in the initiation of the brain masculinization is the hypothalamus [95]. Hypotalamic paraventricular nucleus controls the activity of HPA axis by means of CRH and arginine – vasopressin secretion into pituitary portal system [155]. In adults testosterone can modulate HPA axis by stimulation of expression of the second type receptors to corticotropin-releasing hormone (CRHR 2) in septum [208] and hippocampus [78].

Ventral region of septum has bilateral relationship with the paraventricular nucleus and can therefore play the important role in regulating the activity of HPA axis [159]. The hippocampus is critical in mediating glucocorticoid-dependent negative feedback and thus contributes to control the duration of response to stress [78]. The findings, obtained on the gonadectomized rats with low and high testosterone replacement, indicate that the inhibitory effects of testosterone on corticotrope responses to stress may be linked to reducing in plasma and intrapituitary CBG. This increases the availability of corticosterone to its receptors and enhances glucocorticoid feedback regulation of ACTH release and/or proopiomelanocortin processing [203].

The literature on the role of HPA axis in the regulation of aggressive behavior is rather contradictory. For example, Mikics et al. (2004) showed that a sharp increase in the level of glucocorticoids promotes aggression, and acute blockade of glucocorticoid synthesis reduces aggression between rodent males [123]. According to other sources, the low reactivity HPA axis [202] or chronically low levels of glucocorticoids [91] lead to excessive aggression and even to pathological forms of aggression, including violence. The data of Haller J. et al. (2004) demonstrate that, in rats, chronic but not acute glucocorticoid deficiency induces abnormal attack patterns, deviant cardiovascular responses and social deficits that are similar to those seen in abnormally violent humans [70]. J. Haller and M.R. Kruk (2006) proposed that human aggressiveness is mainly due to three major factors: (i) brain dysfunction affecting aggression-controlling brain centers; (ii) hypoarousal associated with chronically low plasma glucocorticoids, which facilitate violence by diminishing emotional barriers that limit such behaviors; (iii) hyperarousal which leads to irritability and outbursts [71].

According to our data, hyperactivity of HPA axis is manifested by submissive behavior and therefore by predisposition to spontaneous aggression.

Because during fetal development and early neonatal period the thyroid hormones stimulate proliferation, differentiation, migration of neurons and glial cells [162], influence processes of synaptogenesis and myelination of nerve fibers [125], they may be involved to formation of alternative types of behavior.

According to the results of our study, rats with alternative types of behavior did not differ in the content of T_3 in puberty and T_4 in young reproductive age. In other age groups the level of thyroid hormones was higher in dominant animals compared to submissive rats.

According to our data, the dominant rats revealed a high positive correlation between testosterone and thyroxine, which indicates the possibility of a positive influence of testosterone on the synthesis of thyroid hormones in the male with dominant pattern of behavior. There is a close correlation between blood plasma level of T_3 and serotonin content in the hippocampus of young reproductive period males, regardless of the type of behavior. It should be noted that while in the balanced and dominant animals, this relationship is positive in submissive animals it is negative. In dominant rats a correlation between T_3 and serotonin was found not only in the hippocampus but also in the frontal cortex. The correlation between T_3 and serotonin in the frontal cortex was kept in dominant animals of mature reproductive age.

Maybe that in submissive rats of mature reproductive period the process of thyroid hormone synthesis is disturbed, and in the young reproductive period a deiodination process is disturbed. In the available literature the data on the effect of androgens on the deiodinase activity were not found. Nowadays the reducing of the second type 5'-deiodinase activity in depressive disorders is mainly explained by the influence of hypercortisolemia [151]. At the same time, it is impossible to exclude the direct influence of neurotransmitter imbalance, which is observed in affective disorders, on the iodothyronine deiodination process.

In the literature there is evidence of increased activity of 5'-deiodinase type 2 in central nervous system after stimulation of β-adrenergic receptors [66, 131]. It should be noted that in case of 5'-deiodinase type 2 inhibition, a monodeiodination of T_4 is mainly performed by 3 type 5'- deiodinase, resulting in the formation of larger, compared with the norm, amount of rT_3, which itself is a potent inhibitor of 5'-deiodinase type 2 [11].

Correlations between thyroid hormones and serotonin were mainly observed in young reproductive period. If a correlation was observed, between T_3 and serotonin in the balanced and dominant rats it was positive, and in submissive rats it was negative; on the contrary, between T_4 and serotonin in the dominant and balanced males it was negative, in submissive males it was positive.

According to the literature the experimental hypothyroidism in rats is accompanied by a decrease in the concentration of serotonin in the cerebral cortex and mesodiencephalon and by an increase of its circuits in the hippocampus [12, 19, 89]. An administration of thyroid hormone to rats with hypothyroidism and to euthyroid animals leads to an increase of the serotonin concentration in the cortex [67]. Thyroid hormones also cause desensitization autoregulatory 5 - HT1A receptors in the raphe nuclei neurons [67, 129, 189] and increase the density of 5-HT2 receptors in frontal cortex neurons in rats [12, 103].

In turn, serotonin controls the function of the hypothalamic-pituitary-thyroid axis at the level of the hypothalamus, inhibiting the synthesis of thyroid hormones [29, 175].

Analysis of correlations between testosterone, serotonin and T_3 indicates rather the effect of serotonin on the T_3 content of, but not vice versa. As the correlations between the contents of T_3 and noradrenaline are virtually absent and only the correlations between T_4 and noradrenaline are observed, it can be assumed the noradrenaline influence on the secretion of thyroid hormones. This assumption is consistent with the data about the innervation of hypophysiotropic thyrotropin-releasing hormone-synthesizing neurons by catecholamine axons [58].

Monoaminergic brain systems have a great influence on the emotions and behavior in humans and animals [27, 132, 198]. According to our data, in the frontal cortex and the hippocampus of submissive males of all ages, serotonin content was significantly lower, compared to that in the balanced and dominant rats. Obtained results are consistent with most published data on the lack of serotonin neurotransmission in the state of depression [154]. Rat lines with the serotonin content in the frontal cortex and the striatum significantly lower than in the other lines, showed higher levels of anxiety-like behavior and stress reactivity compared with rats with its higher value [68, 107, 218]. It should be noted that both serotonin can affect the response of the HPA axis on stress [79], and glucocorticoids in the brain influence the neurobehavioral functions in animals [105]. For example, altered programming due to maternal stress or glucocorticoid treatment in humans and animals may be associated with HPA axis dysfunction, anxiety and depression [156].

Corticosterone concentrations exhibit a U-shaped dose-response relationship with rates of granule cell death: both complete removal of corticosterone via adrenalectomy and elevated concentrations observed during acute stress result in cell loss. Low basal concentrations are optimal for survival [75]. The observed dependence explains why both abnormally low and high activity of HPA axis is accompanied by abnormal aggression.

The hippocampus plays an integrative role in the regulation of neuroendocrine response to stress, including the psychogenic one. It is critical in mediation of glucocorticoid-dependent negative link and, thus, contributing to the control of response duration to stress [78]. Hippocampus has high level of plasticity and ability to modulation of its functioning. From one side, it is the brain region which is able to neurogenesis in adults. The existence of adult-generated neurons in the human hippocampus was demonstrated by Eriksson et al (1998) [51]. Approximately 6% of the total population of granule cell neurons is generated every month in the dentate gyrus [33]. New neurons have advantages to provide the temporary storage and processing of new memory and to adapt the organism to new environments [211]. From the other side, the functional and structural integration of adult-born neurons into the hippocampal network is influenced by pre-existing astrocytes, synaptic structure and function of the most excitatory synapses by astrocytic perisynaptic processes, which are formed during embryonic or early postnatal development [97].

Rats subjected to chronic unpredictable stress showed a significant elevation of basal plasma corticosterone level, alterations on the mineralocorticod receptor/glucocorticoid receptor mRNA ratio and a decrease in 5-HT 1A receptor mRNA and binding in the hippocampus compared to nonstressed rats [108]. Reduced inhibitory effect of hippocampus on HPA axis causes its hyperactivity [141]. Reduced serotonin content in the hippocampus of submissive rat probably contributes to hyperactivity of HPA axis in these animals.

It should also be noted that the literature on the role of serotonin in the development of aggression is rather contradictory. For example, in some studies, negative correlation was observed [132], while in the others, it was positive [198] between aggressiveness and concentration of the neurotransmitter in the animal brain. To some extent, this contradiction is due to the fact that in the studies of the anti-aggressive effects of the agonists of different types of serotonin receptors, their localization (at the presynaptic or postsynaptic membrane) is often not taken into account. It should be noted that 5-HT-neurotransmission is closely regulated by some autoreceptors, correcting it by inhibitory feedback effects, and localized on either the cell body (mainly 5-HT 1A-autorecdeptors) or on axonal terminals (mainly 5-HT 1B-autoreceptors) [120]. Excess expression of 5-HT 1A-autoreceptors is involved in the reduction of serotoninergic neurotransmission, and it is associated with depression and suicide [4, 148]. de Boer et al. [39] emphasize that specific anti-aggressive effects of the agonists of 5-HT 1A and 5-HT 1B receptors are provided mainly by reducing serotonin transmission in the brain, that is, aggression is associated with activation of serotonin transmission. The different role of serotonin is assumed in adaptive forms of aggression such as social dominance (activation of serotoninergic neurotransmission) versus abnormal forms of aggression (reduced serotoninergic neurotransmission) [132].

Individuals with impulsive aggression and patients with suicidal behavior show a reduction in brain serotoninergic transmission [71]. In addition, impulsive aggression in patients is characterized by the HPA axis hyperactivity. Controlled aggression is characterized by low emotional activity [71, 183]. According to our results, in dominant males the serotonin content was higher in the hippocampus of juvenile animals, as well as in the hippocampus and the frontal cortex in the rats of mature reproductive period (compared with balanced animals of similar age groups). An increase in serotonin content in the rat brain can result from several reasons: increased activity of tryptophan hydroxylase, defect of high affinity serotonin transporter, reduced activity of MAO A.

For tryptophan-hydroxylase activity in aggressive animals, literature data are very contradictory. In particular, in rats demonstrating lack of aggressiveness relative to humans, higher activity of tryptophan hydroxylase-2 and increased serotonin content were found, compared with aggressive rats [100]. In other experiments, decreases in the activity of tryptophan hydroxylase-2 and aggression were found in mice that were the carriers of the mutant gene of this enzyme [101]. $Tph2^{-/-}$ mice exhibited increased depression-like behavior in the forced swim test, decreased anxiety-like behavior and strong aggressiveness observed in the resident–intruder paradigm [127]. Kulikov et al (2005) showed that intensity of mouse intermale aggression was positively associated with activity of the key enzyme of 5-HT synthesis - tryptophan hydroxylase 2 (TPH2) in mouse brain [101]. Administration of tryptophan (300 mg/kg, i.p.) to CC57BR/Mv (low TPH2 activity, low aggressiveness) mice significantly increased the 5-HT and 5-HIAA levels in the midbrain and the number of attacks and their duration in the resident-intruder test. The administration of TPH2 inhibitor to C57BL/6J (high TPH2 activity, high aggressiveness) mice dramatically reduced the 5-HT and 5-HIAA contents in brain structures and attenuated the frequency and the duration of aggressive attacks [102].

High affinity serotonin transporter is important in the regulation of the release and duration of its effect on pre- and post-synaptic receptors in the brain of humans and animals [183]. In knockout mice, by the gene for high affinity serotonin transporter, decreased removal of serotonin from the synaptic cleft was observed, followed by an increase in its concentration in the extracellular fluid in the structures of the forebrain, and a decrease in the autoreceptor function [118]. These neurochemical changes are associated with anxious behavior [35, 56], activation of the HPA axis [3, 35], with development of depression [76]. Deletion of the serotonin transporter gene produces a reduction in aggressive behavior and home cage activity [82]. The decrease in the functioning of a high affinity serotonin transporter at the genetic level is accompanied by the development of depression, and its pharmacological inactivation has anxiolytic and anti-depressant effects [124].

It was also found that knockout mice for MAO A had an increased content of extracellular serotonin in the pre-frontal cortex and hippocampus and showed increased fear behavior and increased non-specific aggression [49, 147]. Very rare point mutation of MAO A, which is accompanied by a decrease in enzyme activity, leads to impulsive aggression also in humans [30]. Despite the fact that both the decrease in the activity of MAO-A, and gene defect of a high affinity serotonin transporter led to increased stress reactivity, increased aggression was observed only in animals deficient in this enzyme, perhaps due to its participation in the metabolism not only of serotonin, but also catecholamines in the brain [83].

We have found a strong negative correlation between the content of testosterone in the blood plasma and serotonin in the brain (hippocampus and frontal cortex) in two age groups of balanced males (in the juvenile and young reproductive period). In balanced males of mature reproductive period, correlation was not observed. In rats of the young and mature reproductive period of dominant and submissive types of behavior, there was a close positive correlation between the content of testosterone in the blood plasma and serotonin in the frontal cortex and the hippocampus. Our results are consistent with the literature data, according to which testosterone, being transformed into estradiol, increased the activity of serotoninergic neurons and the density of 5-HT 2A receptors in the anterior frontal cortex, anterior olfactory cortex and the nucleus accumbens, i.e. in the limbic brain structures [53]. Kubala et al [98] have noted that aggression, in the context of high threat, is triggered by testosterone, and males with low serotonin level are more aggressive in the context of low threat if they are provoked.

It can be assumed that the change in correlation between testosterone and serotonin levels, together with differences in testosterone content, is an important part in forming either submissive or dominant type of behavior.

Norepinephrine-dependent modulation of long-term alterations in synaptic strength, gene transcription and other processes suggest a potentially critical role of this neurotransmitter system in experience-dependent alterations in neural function and behavior [13]. It is possible that norepinephrine is one of the inducers of androgen-dependent sexual brain differentiation [157]. This suggestion is indirectly supported by negative sign of correlation between testosterone and noradrenaline and positive sign of correlation between HPA axis components and noradrenaline in males of all age groups independly on type of behavior.

In all age groups the content of norepinephrine in both the hippocampus and frontal cortex was higher in submissive animals compared to the balanced and dominant ones. In all age groups, with the exception of juvenile period males (hippocampus), the content of norepinephrine in the dominant males was lower compared to the balanced. Some studies demonstrate that increasing of noradrenaline correlates to impulsive aggression, whereas other studies demonstrate an opposite relationship [138]. The noradrenergic system, and particularly its projections to the prefrontal cortex, strongly modulates aggressive behaviors. The increase of density of tyrosine hydroxylase (+) and noradrenaline (+) fibers in prefrontal cortex leads to considerable decrease of aggression in mice genetically predisposed to increased aggression during social interactions [32].

The content of dopamine in submissive males was lower compared with both dominant and balanced rats. The difference between the content of dopamine in the frontal cortex in the balanced and dominant males was observed only in young reproductive period, and in the hippocampus it was revealed only in mature reproductive age. Dopaminergic mesolimbic and mesocortical pathways have projections from the ventral mesencephalon to limbic and cortical structures that regulate cognitive and emotional functions [107]. Mesocorticolimbic dopaminergic and serotoninergic systems modulate state of frontostriatal circles and limbic structures [136].

The changes of monoamine concentrations observed in submissive and dominant rats may be peredetermined by the influence of testosterone on the enzymes of monoamine inactivation. According to literature data, during brain sex differentiation the activity of chatechol O-methyl transferase in the hypothalamus is inhibited by derivatives of 17β-estradiol [157]. This leads to temporary increase of

47

noradrenaline concentration in this structure and masculinization inducing. According to the other results, the sex-dependent differentiation of monoamine oxidase (MAO) in the hypothalamus of 60-day-old, Charles River rats was found. This differentiation involve only type A (MAO-A), and not type B (MAO-B) enzyme [196].

MAO-isoforms have different affinity to monoamines. MAO-A has a high affinity for serotonin and noradrenaline, whereas MAO-B prefers β-phenylethylamine as substrate [172]. The metabolism of other monoamines, such as dopamine, is contributed by both isoforms in most animal species [21]. While dopamine is degraded by MAO A in mice, it is mainly metabolized by MAO B in primates [176]. According to Mattay V.S. et al. (2003), catechol O-methyltransferase, which inactivates released dopamine through enzymatic conversion to 3-methoxytyramine, appears to play a unique role in regulating DA flux in the PFC because of the low abundance and minimal role of DA transporters in the PFC [119]. MAO A is a critical regulator of the homeostatic balance of 5-HT and NA in the brain [24].

Sex-dependent differentiation of MAO-A may be involved in the formation of different types of behavior. This is confirmed by literature data that male MAO-A KO mice show marked reactive aggression and a distinct inability to give the adequate defensive response to different contextual cues [60], mice treated with prolonged administration of MAO-A inhibitors in adulthood do not develop the same abnormalities observed in MAO-A KO mice [24]. The findings M. Bortolato et al. (2012) suggest that MAO A modulates aggressive behaviors by controlling the structure and function of synaptic NMDARs in the PFC [22]. The changes in NMDARs were only found in the PFC [22]. The activation of 5-HT$_{1A}$ receptors has been found to modulate the expression of NMDAR subunits in prefrontal pyramidal neurons cells [215].

It should be noted that the brain monoaminergic system is also involved in the development of anxiety and depression. The most common is monoaminergic theory of depression. According to this theory, the development of depression is associated with abnormalities in serotonin, noradrenaline and dopamine transmission [105]. Frequent detection of excessive aggression and violence in anxiety and depression indicates partial overlaping between neural pathways and neurochemical systems that regulate aggression and anxiety [132].

Published data on the role of monoamines in the anxiety occurrence and the depression development are also contradictory [8, 26, 83, 88, 188]. Analysis of literature data indicates, that serotonin facilitates the conditioned anxiety but inhibits unconditioned fear [63, 135, 200]. The ascending 5-HT pathway that originates in the dorsal raphe nucleus (DRN) and innervates the amygdala and frontal cortex facilitates conditioned fear, while the DRN-periventricular pathway innervating the periventricular and periaqueductal gray matter (PAG) inhibits inborn fight/flight reactions [200]. While activation of 5-HT(1A) or 5-HT(2) receptors in forebrain targets such as the amygdala or hippocampus enhances anxiety-like behaviours in rodents, stimulation of both receptor subtypes in the midbrain PAG markedly reduces anxiety-like behaviour [135]. NA system is involved in the initial alarm reaction [178]. Dopamine plays an important role in fear and anxiety modulating a cortical brake which the medial prefrontal cortex exerts on the anxiogenic output of the amygdala and has an important influence on the trafficking of impulses between the basolateral and central nuclei of amygdala [42].

The main function of fear and anxiety is to act as a signal of danger, threat, or motivational conflict, and to trigger appropriate adaptive responses [178]. Some authers undistinguishe fear and anxiety. According to other, the object of fear is real and the origins of anxiety are unclear [178]. Key components of fear circuitry include the amygdala (and its subnuclei), nucleus accumbens (including BNST), hippocampus, ventromedial hypothalamus, periaqueductal gray, a number of brain stem nuclei, thalamic nuclei, insular cortex, and some prefrontal regions (mainly infralimbic cortex) [174]. In all mammalian species, there are three distinct sites in the brain where electrical stimulation will provoke a full fear response: the lateral and central zones of the amygdala, the anterior and medial hypothalamus, and specific areas of the PAG [178]. The amygdala and the medial prefrontal cortex are involved in the extinction of conditioned fear. A defined subpopulation of basal amygdala projection neurons targeting the prelimbic subdivision of mPFC is active during states of high fear, whereas basal amygdala neurons targeting the infralimbic subdivision of mPFC are recruited during fear extinction [169]. Convergent inputs from both the ventral hippocampus and prelimbic prefrontal cortex in the amygdaloid basal nuclei mediate the contextual control of fear after extinction [139].

BLA is important for processing both positive and negative affect, the glutamatergic pathway from the BLA to the nucleus accumbens, in conjunction with dopamine signalling in the the nucleus accumbens, promotes motivated behavioural responding [182].

Brain areas relevant for the control of aggression include cortex, amygdala, septum, hypothalamus, periaqueductal grey and the locus coeruleus [69]. The lateral hypothalamus and central amygdala are tightly involved in predatory aggression [195].

The individual typological characteristics of the organism are based on the activity of four motivational emotiogenic brain structures (frontal cortex, hippocampus, amygdala and hypothalamus). Lateral hypothalamus is positive emotiogenic structure [116, 216] and medial hypothalamus [116], tegmentum of the midbrain [216] belong to emotionally negative brain structures.

Virtually all of noradrenergic fibers begin in the nuclei of the brain stem. Almost half of them belong to the locus coeruleus (LC) [150]. The LC has dense excitatory projections to the cortex, the hippocampus, the amygdale, the thalamus, the hypothalamus, serotoninergic neurons of the dorsal raphe nucleus (DRN) [163]. Despite the wide noradrenergic projection from LC to the cortex, mPFC is the only cortical structure projecting back to the LC [109]. The exposure to stress is associated with the LC excitation and the increase of noradrenaline releasing and renovation in brain areas having noradrenergic innervation [26]. The LC is believed to be the sole source of noradrenaline to cortex and hippocampus [163]. But Robertson S.D. et al (2013) have identified an unexpected projection to the prefrontal cortex, challenging the long-held belief that the locus coeruleus is the sole source of norepinephrine projections to the cortex [160].

Noradrenergic projections to cortex, hippocampus, amygdala (particularly to the central and basal nuclei), DRN, hypothalamic paraventricular nucleus are rather excitatory but to lateral hypothalamus area they are inhibitory [163], however excitatory α_1-adrenoreceptors are also identified in the lateral hypothalamus, where activation of these receptors is linked to behavioral activation and exploration [180].

The exposure to stress is accompanied by stimulation of DRN by both noradrenergic projections and by CRH. In the medial prefrontal cortex (mPFC), a brain area implicated in mood and anxiety disorders and a major target of ascending 5-HT pathways, studies have shown an increase in 5-HT or 5-HT metabolites after administration of CRH [122]. A majority of pyramidal neurons in the mPFC are inhibited by 5-HT. Possible mechanisms for the inhibitory effects of 5-HT on mPFC pyramidal neurons include direct effects, mediated by 5-HT1A receptors, or indirect effect, mediated through activation of GABAergic interneurons via 5-HT2A and /or 5-HT3 receptors [122]. This leads to the inhibition by 5-HT of the descending excitatory drive from pyramidal neurons in the mPFC to limbic areas such as the extended amygdala, e.g. the central nucleus of the amygdala and BNST. There is strong evidence to suggest that the input from the ventral mPFC to the central nucleus of the amygdala is an important axis for controlling the extinction of conditioned fear [122] and BNST plays the important role in the expression of unconditioned anxiety-like behaviors [206]. It should be noted, that pyramidal neurons in the rat PFC that simultaneously project to the ventral tegmental area and DRN express 5-HT2A (excitatory) receptors [36].

Dopaminergic neurons in the ventral tegmental area (VTA) receive excitatory input from and send reciprocal projections to the mPFC. Glucocorticoids act locally within the mPFC to modulate mesocortical dopamine efflux by potentiation of glutamatergic drive onto dopaminergic neurons in the VTA [31]. The ventral tegmental area has dopaminergic projections to both lateral hypothalamus and amygdala [170]. Dopaminergic system forms hedonistic behavior components (pleasure, satisfaction and aspiration for them).

Corticotropin-releasing factor regulates instinctive forms of emotional behavior (fear, anxiety, frustration and deliverance from them). The system of extended amygdala is considered to be the basis of extrahypothalamic CRF, influencing the stress-dependent behavior, initiating emotionally motivated responses and mediating anxiogenic effects of CRH [170]. The limbic-hypothalamic-pituitary-adrenal axis encompasses inputs from the amygdala and prefrontal cortex to the hippocampus, which inhibits the PVN of the hypothalamus. An indirect pathway involves excitatory input from the hippocampus to the bed nucleus of the stria terminalis, which provides the inhibition of the PVN [75].

Fear or anxiety result in the expression of a range of adaptive or defensive behaviors, which are aimed at escaping from the source of danger or motivational conflict [178]. The PFC is thought to participate in high-level control of the generation of behaviors (including the decision to execute actions). The prefrontal cortex efferent projections to limbic areas facilitate a top-down control on the execution of goal-directed behaviours [43]. The selective activation of mPFC cells projecting to the brainstem DRN induces a rapid and reversible effect on selection of the active behavioral state [207]. Alterations of the orbitofrontal cortex have been shown to lead to ineffective allocation of adaptive strategies and deficits in executive functions [114]. According to W.B. Hoover et al. (2007), the infralimbic division of the mPFC in rats appears to represent a visceromotor center homologous to the orbitomedial PFC of primates [84]. Pyramidal neurons in the PFC integrate multiple synaptic signals from different brain areas, and project to main components of the limbic-subcortical circuit that regulate negative affect and reactive aggression, such as the amygdaloid nuclei, medial hypothalamus and dorsal periaqueductal gray [22].

Anatomical and immunohistochemical studies have shown that the cerebellum is extensively innervated by both 5-HT and noradrenaline fibers [18] and they modulate the firing rate of Purkinje cells and cerebellar nuclei [130]. Serotonin is critical for cerebellar development and its normal function in the mature state [5]. During early development, the cerebellum exhibits a transient expression of 5-HT1 and 5-HT3 receptors, which play an important role in the formation of the cerebellar neuronal network [137]. The cerebellum exhibits postnatal neurogenesis [20]. Serotonin increases cell proliferation and differentiation of cerebellar neural progenitor cells at low concentrations, but administration of high concentrations of 5-HT significantly decreases cerebellar cell proliferation [220]. Over activation of the noradrenaline system in early life by the β_2 agonist terbutaline leads to degeneration of Purkinje cells and a thinning of the molecular and granular layers [158]. Noradrenaline can modulate neuronal circuitry within the cerebellum through a direct inhibitory effect on Purkinje cells [213]. Taking into account the neural connections of the cerebellum with the cerebral cortex and limbic system structures, the pointedness of changes in the content of catecholamines in the cerebellum, frontal cortex and hippocampus, the existence of correlations between the content of monoamines in the cerebellum and structures of emotiogenic limbiconeocortical system, it can be assumed that the cerebellum is involved in the formation of a dominant/submissive behaviors, in the development of aggressive and depressive states.

Analysis of own results and literature data makes it possible to make the assumption that biogenic monoamines are the major players in the development both depression and aggression. First of all, it is possible that namely noradrenaline is one of the inducers of androgen-dependent sexual brain differentiation [157]. Serotonin and noradrenaline are known to exert a profound influence on early brain development through the regulation of neurogenesis, migration, differentiation, plasticity and other key morphogenetic processes [190].

The dependence of male behavior type on testosterone level was revealed in our study. Hyperactivity of HPA system was found in submissive animals. The presence of tight negative correlation between testosterone level and HPA system hormones (ACTH and corticosterone) was shown in male rat indepndly of age and behavior type, which is evidence of testosterone participation in HPA system organization and inhibition influence of testosterone on this system.

Imbalance in testosterone/serotonin and testosterone/cortisol ratios increases the predisposition to aggression because of reduced activation of the neural circuitry of impulse control and self-regulation [143]. According to our results, in the dominant rats the level of testosterone was higher and the level of corticosterone was lower, than in balanced ones. In the submissive rats the testosterone level and serotonin content were lower, but corticosterone level was higher, than in the balanced ones. Serotonin facilitates prefrontal inhibition, and thus insufficient serotonergic activity can enhance aggression [143].

Initially we speculated that aggression is finally due to the imbalance between positive and negative emotions [146]. Fear and anxiety belong to negative emotions. Monoamines are the major mediator systems involved in the formation of fear and anxiety. Serotonin facilitates conditioned fear and inhibits unconditioned fear [62, 122]. Norepinephrine intiates unconditioned fear, CRH release [178], lateral hypothalamus stimulation, it mediates positively motivated exploratory and approach activities [180]. We supposed that the decrease in the concentration of serotonin plays an important role in the development of impulsive aggression by reducing the inhibitory effect on the negative emotiogenic structures. The reduction of noradrenaline in the dominant males plays an important role in the development of adaptive aggression by reducing the influence on positive emotiogenic structures. The decrease of dopamine content promotes adaptive aggression in dominant rats and the formation of submissive behavior in submissive males by reducing the mediation of positive emotional reactions.

But the analysis of literature data obtained on MAO- knockout animals, allowed us to conclude that the relationship between biogenic amines is more important, than their absolute amount.

Monoamine oxidase A hypomorphic mice (MAO-A^{Neo}) MAO-A^{Neo} mice showed undetectable MAO-A enzymatic activity in the hippocampus and midbrain, that was accompanied by high levels of 5-HT and NA, and low MAO-A activity in the prefrontal cortex and amygdala that was accompanied by normal 5-HT levels and high NA levels [5]. Noradrenaline level was significantly higher in both the prefrontal cortex and amygdala as compared with wild type mice, but noradrenaline content in amygdala was significantly lower as compared with MAO-A KO mice) [24]. Whereas MAO-A^{Neo} and KO mice showed significant reductions in social interaction, only MAO-A KO mice showed the increase of resident–intruder aggression [24]. MAO B KO mice revealed a significant elevation in whole-brain levels of β-phenylethylamine, but not other monoamines [65]. The most marked increase of β-phenylethylamine levels were observed in striatum and prefrontal cortex, which are implicated in behavioral disinhibition [210]. MAO B deficiency may result in behavioral disinhibition and negligence of potential or actual dangers [86]. The involvement of striatum and prefrontal cortex in behavioral disinhibition has been linked to the functional activity of DAergic system [142]. Dopamine plays the key role in behavioral disinhibition [199] and anxiolysis [144]. β-Phenylethylamine induces modification of the DA signaling, which may play a role in the behavioral responses mediated by β-phenylethylamine [23]. Experimental evidence indicates that DA influences PFC function in accordance with an inverted U-shaped dose–response curve, such that the response is optimized within a narrow range of DA activity [119].

The calculations of NA/5-HT ratio revealed its increase (from 3.3 to 8.3 times depending on brain region and age, $P<0.05$) in submissive rats and decrease (from 1.54 to 2.9 times depending on brain region and age, $P<0.05$) in dominant rats as compared with balanced ones. NA/DA ratio was increased (from 4.4 to 7.6 times depending on brain region and age, $P<0.05$) in submissive rats versus balanced males. In dominant rats statistically significant decrease of NA/DA ratio was observed only in the hippocampus (2.1 times, $P<0.05$) and cerebellum (1.7, $P<0.05$) in mature reproductive period.

Thus, both submissive and dominant males are characterized by imbalance between monoamines:

• The significant increase of both NA/5-HT and NA/DA ratios in all investigated brain regions was observed in submissive rats of all age groups.

• The significant decrease of NA/5-HT ratio was revealed in all investigated brain regions in dominant rats of all age groups.

This imbalance may be caused by changes of the proportion between enzymes, involving in degradation of monoamines, in early brain development and can provide predisposition to different types of aggression.

Disclosure of Conflict of Interests: The authors state that they have no conflict of interests.

REFERENCES

1.   aan het Rot M, Mathew SJ, Charney DS. Neurobiological mechanisms in major depressive disorder. CMAJ. 2009;180(3):305-13. doi: 10.1503/cmaj.080697. Cited in PubMed; PMID 19188629.

2.   Abdelgadir SE, Resko JA, Ojeda SR, Lephart ED, McPhaul MJ, Roselli CE. Androgens regulate aromatase cytochrome P450 messenger ribonucleic acid in rat brain. Endocrinology. 1994;135(1):395–401. Cited in PubMed; PMID 8013375.

3.   Adamec R, Burton P, Blundell J, Murphy DL, Holmes A. Vulnerability to mild predator stress in serotonin transporter knockout mice. Behav Brain Res. 2006;170(1):126-40. Cited in PubMed; PMID 16546269.

4.   Albert PR, Le François B, Millar AM. Transcriptional dysregulation of 5-HT1A autoreceptors in mental illness. Mol Brain. 2011;4:21. doi: 10.1186/1756-6606-4-21. Cited in PubMed; PMID 21619616.

5.   Alzghoul L, Bortolato M, Delis F, Thanos PK, Darling RD, Godar SC, Zhang J, Grant S, Wang GJ, Simpson KL, Chen K, Volkow ND, Lin RC, Shih JC. Altered cerebellar organization and function in monoamine oxidase A hypomorphic mice. Neuropharmacology. 2012;63(7):1208-17. doi: 10.1016/j.neuropharm.2012.08.003. Cited in PubMed; PMID 22971542.

6.   Ambrogin P, Cuppini R, Ferri P, Mancini C, Ciaroni S, Voci A, Gerdoni E, Gallo G. Thyroid hormones affect neurogenesis in the dentate gyrus of adult rat. Neuroendocrinology. 2005;81(4):244-53. Cited in PubMed; PMID 16113586.

7.   Amore M. Partial androden deficiency and neuropsychiatric symptoms in aging men. J. Endocrinol invest. 2005;28:49-54. Cited in PubMed; PMID 16760626.

8.   Avgustinovich DF, Alekseyenko OV, Bakshtanovkaya Iv, Koryakina LA, Lipina NV, Tenditnik MV, Bondar NP, Kovalenko IL, Kudryavtseva NN. Dynamic changes of brain serotonergic and dopaminergic activities during development of anxious depression: experimental study. Usp Fiziol Nauk. 2004;35(4):19-40 Cited in PubMed; PMID 15573884.

9.   Bale TL, Vale WW. CRF1 and CRF2 receptors: role in stress responsivity and other behaviors. Annu Rev Pharmacol Toxicol. 2004;44:525–57. Cited in PubMed; PMID 14744257.

10.   Barden N. Implication of the hypothalamic-pituitary-adrenal axis in the physiopathology of depression. J. Psychiatry Neurosci. 2004; 29: 185–93. Cited in PubMed; PMID 15173895.

11.   Bates JM, St Germain DL, Galton VA. Expression profiles of the three iodothyronine deidinase, D1, D2 and D3, in the developing rat. Endocrinology. 1999;140(2):844-51. Cited in PubMed; PMID 9927314.

12.   Bauer M., Szuba M.P., Whybrow P.C. Psychiatric and behavioral manifestation of hyperthyroidism and hypothyroidism. Psychoneuroendocrinology: the scientific basis of clinical practice. Ed. by O.M. Wolkowitz, A.J. Rothschild – Washington: American Psychiatric Publishing, Inc., 2003. P. 419-444.

13.   Berridge CW, Waterhouse BD. The locus coeruleus-noradrenergic system: modulation of behavioral state and state-dependent cognitive processes. Brain Res Brain Res Rev. 2003;42(1):33-84. Cited in PubMed; PMID 12668290.

14.   Bevzyuk D.A. Neurobiological features of behavior of rats-aggressors in terms of influence on the emotional structure of the hypothalamus and central gray matter. Experimental and clinical medicine. 2006;(3):42-5.

15.   Bingaman EW, Van de Kar LD, Yracheta JM, Li Q, Gray TS. Castration attenuates prolactin response but potentiates ACTH response to conditioned stress in the rat.Am J Physiol. 1995;269(4Pt2):R856-63. Cited in PubMed; PMID 7485603.

16.   Bingaman EW, Magnuson DJ, Gray TS, Handa RJ. Androgen inhibits the increases in hypothalamic corticotropin-releasing hormone (CRH) and CRH-immunoreactivity following gonadectomy. Neuroendocrinology. 1994;59:228–34. Cited in PubMed; PMID 8159272.

17.   Bingham B, Gray M, Sun T, Viau V. Postnatal blockade of androgen receptors or aromatase impair the expression of stress hypothalamic-pituitary-adrenal axis habituation in adult male rats. Psychoneuroendocrinology. 2011;36(2):249-57. doi: 10.1016/j.psyneuen. 2010.07.015. Cited in PubMed; PMID 20719434.

18.   Bishop GA, Ho RH. The distribution and origin of serotonin immunoreactivity in the rat cerebellum. Brain Res. 1985;331:195–207. Cited in PubMed; PMID 986565.

19.   Bjerke SN, Bjøro T, Heyerdahl S. Psychiatric and cognitive aspects of hypothyroidism. Tidsskr Nor Laegeforen. 2001;121(20):2373-6. Cited in PubMed; PMID 11603044.

20.   Bonfanti L, Ponti G. Adult mammalian neurogenesis and the New Zealand white rabbit. Vet J. 2008;175(3):310-31. Cited in PubMed; PMID 17391998.

21.   Bortolato M, Chen K, Shih JC. Monoamine oxidase inactivation: from pathophysiology to therapeutics. Adv Drug Deliv Rev. 2008;60:1527–1533. doi: 10.1016/j.Cited in PubMed; PMID 18652859.

22. Bortolato M, Godar S C, Melis M, Soggiu A, Roncada P, Casu A, Flore G, Chen K, Frau R, Urbani A, Castelli M P, Devoto P, and Shih J C. NMDARs mediate the role of monoamine oxidase A in pathological aggression. J Neurosci. 2012; 32(25): 8574–8582. doi: 10.1523/JNEUROSCI.0225-12.2012. Cited in PubMed; PMID 22723698.

23. Bortolato M, Godar SC, Davarian S, Chen K, Shih JC. Behavioral disinhibition and reduced anxiety-like behaviors in monoamine oxidase B-deficient mice. Neuropsychopharmacology. 2009;34(13):2746-57. doi: 10.1038/npp.2009.118. Cited in PubMed; PMID 19710633.

24. Bortolato M, Shih JC. Behavioral outcomes of monoamine oxidase deficiency: preclinical and clinical evidence. Int Rev Neurobiol. 2011;100:13-42. doi: 10.1016/B978-0-12-386467-3.00002-9. Cited in PubMed; PMID 21971001.

25. Brede M, Nagy G, Philipp M, Sorensen JB, Lohse MJ, Hein L. Differential control of adrenal and sympathetic catecholamine release by alpha 2-adrenoceptor subtypes. Mol Endocrinol. 2003; 17(8):1640-46. Cited in PubMed; PMID 12764077.

26. Bremner JD, Krystal JH, Southwick SM, Charney DS. Noradrenergic mechanisms in stress and anxiety: Preclinical studies. Synapse. 1996; 23(1): 28-38. Cited in PubMed; PMID 8723133.

27. Briley M, Moret C. Improvement of social adaptation in depression with serotonin and norepinephrine reuptake inhibitors. Neuropsychiatr Dis Treat. 2010;6:647-55. doi: 10.2147/NDT.S13171. Cited in PubMed; PMID 20957125.

28. Broedel O. Eravci M, Fuxius S, Smolarz T, Jeitner A, Grau H, Stoltenburg-Didinger G, Plueckhan H, Meinhold H, Baumgartner A. Effect of hyper and hypothyroidism on thyroid hormone concentrations in regions of the rat brain. Am J Physiol Endocrinol Metab. 2003;285(3):E470-80. Cited in PubMed; PMID 12736158.

29. Brown S.L. The pathogenesis of depression: reconcideration of neurotransmitter data // Handbook of Depression and Anxiety / S.L. Brown, R.L. Steinberg, H.M. Van Praag // Eds J.A. Den Boer, J.A.Ad Sitsen. N4: M.Dekker, 1994:317-47.

30. Brunner HG, Nelen M, Breakefield XO, Ropers HH, van Oost BA. Abnormal behavior associated with a point mutation in the structural gene for monoamine oxidase A. Science. 1993;262(5133):578-80. Cited in PubMed; PMID 8211186.

31. Butts KA, Phillips AG. Glucocorticoid receptors in the prefrontal cortex regulate dopamine efflux to stress via descending glutamatergic feedback to the

ventral tegmental area. Int J Neuropsychopharmacol. 2013;16(8):1799-807. doi: 10.1017/S1461145713000187. Cited in PubMed; PMID: 23590841.

32.   Cambon   K, Dos-Santos   Coura   R, Groc   L, Carbon   A, Weissmann D, Changeux JP, Pujol JF, Granon S. Aggressive behavior during social interaction in mice is controlled by the modulation of tyrosine hydroxylase expression in the prefrontal   cortex.   Neuroscience.   2010;   171(3):   840-51.   doi: 10.1016/j.neuroscience.2010.09.015. Cited in PubMed; PMID 20923695.

33.   Cameron HA, McKay RD. Adult neurogenesis produces a large pool of new granule cells in the dentate gyrus. J Comp Neurol. 2001;435(4):406-17. Cited in PubMed; PMID: 11406822.

34.   Campeau S, Watson SJ. Neuroendocrine and behavioral responses and brain pattern of c-fos induction associated with audiogenic stress. Neuroendocrinol. 1997;9(8):577–88. Cited in PubMed; PMID 9283046.

35.   Carroll JC, Boyce-Rustay JM, Millstein R, Yang R, Wiedholz LM, Murphy DL, Holmes A. Effects of mild early life stress on abnormal emotion-related behaviors in 5-HTT knockout mice. Behav Genet. 2007;37(1):214-22. Cited in PubMed; PMID 17177116.

36.   Celada P, Puig MV, Artigas F. Serotonin modulation of cortical neurons and networks. Front. Integr. Neurosci. 2013;7:25. doi: 10.3389/fnint.2013.00025. Cited in PubMed; PMID 23626526. PMCID:PMC3630391.

37.   Chichinadze KN, Dominiadze TR, Matitaischvili T, Chichinadze NK, Lazaraschvili AG. Is the blood plasma testosterone level linked with aggressive behaviour of prisoner men? Bul Exp Biol Med. 2010;149(1): 11-3.

38.   Cooke   B,   Hegstrom   CD,   Villeneuve   LS,   Breedlove   SM.   Sexual differentiation   of   the   vertebrate   brain: principles   and   mechanisms.   Front Neuroendocrinol. 1998;19:253–86. Cited in PubMed; PMID 9799588.

39.   de Boer SF, Koolhaas JM. 5-HT1A and 5-HT1B receptor agonists and aggression: a pharmacological challenge of the serotonin deficiency hypothesis. Eur J Pharmacol. 2005;526(1-3):125-39. Cited in PubMed; PMID 16310183.

40.   de Diego AM, Gandía L, García AG. A physiological view of the central and peripheral mechanisms that regulate the release of catecholamines at the adrenal medulla. Acta Physiol (Oxf). – 2008; 192(2): 287-301. Cited in PubMed; PMID 18005392.

41.   de Kloet ER, Joëls M, Holsboer F. Stress and the brain: from adaptation to disease. Nat Rev Neurosci. 2005;(6):463-73.

42.   de la Mora MP, Gallegos-Cari A, Arizmendi-García Y, Marcellino D, Fuxe K. Role of dopamine receptor mechanisms in the amygdaloid modulation of fear and anxiety: Structural and functional analysis. Prog Neurobiol. 2010;90(2):198-216. doi: 10.1016/j.pneurobio.2009.10.010. Cited in PubMed; PMID: 19853006.

43.   Del Arco A, Mora F. Neurotransmitters and prefrontal cortex-limbic system interactions: implications for plasticity and psychiatric disorders. J Neural Transm. 2009;116(8):941-52. doi: 10.1007/s00702-009-0243-8. Cited in PubMed; PMID 19475335.

44.   Delville Y, Mansour KM, Ferris CF. Testosterone facilitates aggression by modulating vasopressin receptors in the hypothalamus. Physiol. Behav. 1996;60(1):25-9. Cited in PubMed; PMID 8804638.

45.   Dennis RL, Chen ZQ, Cheng HW. Serotonergic mediation of aggression in high and low aggressive chicken strains. Poult Sci. 2008;87(4):612-20. Cited in PubMed; PMID 18339980.

46.   Di Matteo V, De Blasi A, Di Giulio C, Esposito E. Role of 5-HT (2c) – receptors in the control of central dopamine function. Trends. Pharmacol. Sci. 2001;22(5): 229-32. Cited in PubMed; PMID 11339973.

47.   Dranovsky A, Hen R. Hippocampal neurogenesis: regulation by stress and antidepressants. Biol Psychiatry. 2006;59(12):1136-43. Cited in PubMed; PMID 16797263.

48.   Drevets WC. Neuroimaging and neuropathological studies of depression: implications for the cognitive-emotional features of mood disorders. Curr Opin Neurobiol. 2001;11:240–49. doi: 10.1016/S0959-4388(00)00203-8. Cited in PubMed; PMID 11301246.

49.   Dubrovina NI, Popova NK, Gilinskii MA, Tomilenko RA, Seif I. Acquisition and extinction of a conditioned passive avoidance reflex in mice with genetic knockout of monoamine oxidase A. Neurosci Behav Physiol. 2006;36(4):335-9. Cited in PubMed; PMID 16583159.

50.   Duman RS, Heninger GR, Nestler EJ. A molecular and cellular theory of depression. Arch Gen Psychiatry.1997;54(7):597–606. Cited in PubMed; PMID 9236543.

51.   Eriksson PS, Perfilieva E, Björk-Eriksson T, Alborn AM, Nordborg C, Peterson DA, Gage FH. Neurogenesis in the adult human hippocampus. Nat Med. 1998;4(11):1313-7. Cited in PubMed; PMID: 9809557.

52.   Fahim C, He Y, Yoon U, Chen J, Evans A, Pérusse D. Neuroanatomy of childhood disruptive behavior disorders. Aggress behav. 2011;37(4):326-37. Cited in PubMed; PMID 21538379.

53.   Ferris CF, Delville Y. Vasopressin and serotonin interactions in the control of agonistic behavior. Psychoneuroendocrinology. 1994;19(5-7):593-601. Cited in PubMed; PMID 7938357.

54.   Fink G. Sumner BE, Rosie R. Estrogen control of central neurotransmission: effect on mood, mental state, and memory. Cell Mol Neurobiol. 1996;16(3):325-44.

55.   Foote SL, Morrison JH. Extrathalamic modulation of cortical function. Annu. Rev. Neurosci. 1987; 10: 67-95. Cited in PubMed; PMID 3551766.

56.   Fox MA, Andrews AM, Wendland JR, Lesch KP, Holmes A, Murphy DL. A pharmacological analysis of mice with a targeted disruption of the serotonin transporter. Psychopharmacology(Berl). 2007;195(2):147-66. Cited in PubMed; PMID 17712549.

57.   Frye CA, Seliga AM. Testosterone increases analgesia, anxiolysis, and cognitive performance of male rats. Cogn. Affect. Behar. Neurosci. 2001;1: 371-81. Cited in PubMed; PMID 12467088.

58.   Füzesi T, Wittmann G, Lechan RM, Liposits Z, Fekete C. Noradrenergic innervation of hypophysiotropic thyrotropin-releasing hormone-synthesizing neurons in rats. Brain Res. 2009;1294:38-44. doi: 10.1016/j.brainres.2009.07.094. Cited in PubMed; PMID 19651110.

59.   Gelinas S, Bureau G, Valsatro B, Massicotte G, Cicchetti F, Chiasson K. α-And β-estradiol protect neuronal but not native PC12 cells from paraquat-induced oxidative stress. Neurotox Res. 2004; 6: 141–48. Cited in PubMed; PMID15325966.

60.   Godar SC, Bortolato M, Frau R, Dousti M, Chen K, Shih JC. Maladaptive defensive behaviours in monoamine oxidase A-deficient mice. Int J Neuropsychopharmacol. 2010;15:1–13. doi: 10.1017/S1461145710001483. Cited in PubMed; PMID 21156093.

61.   Gonçalves D, Saraiva J, Teles M, Teodósio R, Canário AV, Oliveira RF. Brain aromatase mRNA expression in two populations of peacock blenny Salaria pave with divergent mating systems. Horm Behav. 2010;57(2):155-61. Cited in PubMed; PMID 19840804.

62.   Graeff FG, Guimarães FS, De Andrade TG, Deakin JF. Role of 5-HT in stress, anxiety, and depression. Pharmacol Biochem Behav. 1996;54(1):129-41. Cited in PubMed; PMID: 8728550.

63.   Graeff FG. Serotonin, the periaqueductal gray and panic.Neurosci Biobehav Rev. 2004;28(3):239-59. Cited in PubMed; PMID: 15225969.

64.   Grgurevic N, Büdefeld T, Tobet AS, Rissman EF, Majdic G. Aggressive Behaviors in Adult SF-1 Knockout Mice That Are Not Exposed to Gonadal Steroids During Development. Behav Neurosci. 2008; 122(4): 876–84. Cited in PubMed; PMID 18729641.

65.   Grimsby J, Toth M, Chen K, Kumazawa T, Klaidman L, Adams JD, et al. Increased stress response and beta-phenylethylamine in MAOB-deficient mice. Nat Genet. 1997;17(2):206–210. Cited in PubMed; PMID 9326944.

66.   Gur E, Lifschytz T, Lerer B, Newman ME. Effects of triodothyronine and imipramine on basal 5-HT levels and 5-HT(1) autoreceptors activity in rat cortex. Eur J Pharmacol. 2002;457(1):37-43. Cited in PubMed; PMID 12460641.

67.   Haas MJ, Mreyoud A, Fishman M, Mooradian AD. Microarray analysis of thyroid hormones –induced changes in mRNA expression in the adult rat brain. Neurosci Lett. 2004;365(1):14-8. Cited in PubMed; PMID 15234464.

68.   Hackler EA, Airey DC, Shannon CC, Sodhi MS, Sanders-Bush E. 5-HT(2C) receptor RNA editing in the amygdala of C57BL/6J, DBA/2J, and BALB/cJ mice. Neurosci Res. 2006;55(1):96-104. Cited in PubMed; PMID 16580757.

69.   Halász J, Liposits Z, Kruk MR, Haller J. Neural background of glucocorticoid dysfunction-induced abnormal aggression in rats: involvement of fear- and stress-related structures. Eur J Neurosci. 2002;15(3):561-9. Cited in PubMed; PMID: 11876784.

70.   Haller J, Halász J, Mikics E, Kruk MR. Chronic glucocorticoid deficiency-induced abnormal aggression, autonomic hypoarousal, and social deficit in rats. J Neuroendocrinol. 2004;16(6):550-7. Cited in PubMed; PMID 15189330.

71.   Haller J, Kruk MR. Normal and abnormal aggression: human disorders and novel laboratory models. Neurosci Biobehav Rev. 2006;30(3):292-303. Cited in PubMed; PMID 16483889.

72.   Haller J, Bakos N, Rodriguiz RM, Caron MG, Wetsel WC, Liposits Z. Behavioral responses to social stress in noradrenaline transporter knockout mice: effects on social behavior and depression. Brain Res Bull. 2002;58(3):279-84. Cited in PubMed; PMID 12128153.

73.   Hamelink C, Tjurmina O, Damadzic R, Young WS, Weihe E, Lee HW, Eiden LE. Pituitary adenylate cyclase-activating polypeptide is a sympathoadrenal neurotransmitter involved in catecholamine regulation and

glucohomeostasis. Proc Nat Acad Sci U S A. 2002; 99(1): 461-66. Cited in PubMed; PMID 11756684.

74.   Handa RJ, Nunley KM, Lorens SA, Louie JP, McGivern RF, Bollnow MR. Androgen regulation of adrenocorticotropin and corticosterone secretion in the male rat following novelty and foot shock stressors. Physiol. Behav. 1994;55(1):117–24. Cited in PubMed; PMID 8140154.

75.   Hanson ND, Owens MJ, and Nemeroff CB. Depression, Antidepressants, and Neurogenesis: A Critical Reappraisal Neuropsychopharmacology. 2011; 36(13): 2589–602. doi: 10.1038/npp.2011.220. Cited in PubMed; PMID 21937982.

76.   Hariri AR, Holmes A. Genetics of emotional regulation: the role of the serotonin transporter in neural function. Trends Cogn Sci. 2006;10(4):182-91. Cited in PubMed; PMID 16530463.

77.   Hashimoto H. PACAP is implicated in the stress axes. Curr Pharm Des. 2011; 17(10): 985-9. Cited in PubMed; PMID 21524255.

78.   Herman JP, Mueller NK. Role of the ventral subiculum in stress integration. Behav Brain Res. 2006;174(2):215-24. Cited in PubMed; PMID 16876265.

79.   Herman JP, Ostrander MM, Mueller NK, Figueiredo H. Limbic system mechanisms of stress regulation: hypothalamo-pituitary-adrenocortical axis. Prog Neuropsychopharmacol Biol Psychiatry. 2005;29(8):1201-13. Cited in PubMed; PMID 16271821.

80.   Heuer H, Mason CA. Thyroid hormone induces cerebellar Purkinje cell dendritic development via the thyroid hormone receptor. J Neurosci. 2003;23(33):10604-12. Cited in PubMed; PMID 14627645.

81.   Hodel A. Effects of glucocorticoids on adrenal chromaffin cells. J Neuroendocrinol. 2001 Feb;13(2):216-20. Cited in PubMed; PMID 11168848.

82.   Holmes A, Murphy DL, Crawley JN. Reduced aggression in mice lacking the serotonin transporter. Psychopharmacology (Berl). 2002;161(2):160-7. Cited in PubMed; PMID 11981596.

83.   Holmes A. Genetic variation in cortico – amygdale serotonin function and risk for stress – related disease. Neursci. Biobehav. Rev. 2008;32(7): 1293-314. doi: 10.1016/j.neubiorev.2008.03.006. Cited in PubMed; PMID 18439676.

84. Hoover WB, Vertes RP. Anatomical analysis of afferent projections to the medial prefrontal cortex in the rat. Brain Struct Funct. 2007;212(2):149-79. Cited in PubMed; PMID 17717690.

85.   Hutchison JB Gender-specific steroid metabolism in neural differentiation. Cell Mol Neurobiol. 1997;17(6):603-26. Cited in PubMed; PMID 9442349.

86. Iacono WG, Malone SM, McGue M. Substance use disorders, externalizing psychopathology, and P300 event-related potential amplitude. Int J Psychophysiol. 2003;48 (2):147–78. Cited in PubMed; PMID 12763572.

87. Ito M. Long-term depression. Annu Rev Neurosci. 1989;12:85-102. Cited in PubMed; PMID 2648961.

88. Itoi K[1], Sugimoto N, Suzuki S, Sawada K, Das G, Uchida K, Fuse T, Ohara S, Kobayashi K. Targeting of locus ceruleus noradrenergic neurons expressing human interleukin-2 receptor α-subunit in transgenic mice by a recombinant immunotoxin anti-Tac(Fv)-PE38: a study for exploring noradrenergic influence upon anxiety-like and depression-like behaviors. J Neurosci. 2011;31(16): 6132-9. doi: 10.1523/JNEUROSCI.5188-10.2011. Cited in PubMed; PMID 21508238.

89. Joffe RT, Marriott M. Thyroid hormone levels and recurrence of major depression. Am J Psychiatry. 2000;157(10):1689-91.Cited in PubMed; PMID 11007728.

90. Jorgensen HS. Studies on the neuroendocrine role of serotonin. Dan Med Bull. 2007; 54(4): 266-88. Cited in PubMed; PMID 18208678.

91. Kim JJ, Haller J. Glucocorticoid hyper- and hypofunction: stress effects on cognition and aggression. Ann. N. Y. Acad. Sci. 2007;1113: 291–303. Cited in PubMed; PMID 17513462.

92. Kirkegaard C, Faber J. Free thyroxine and 3,3', 5' – triiodothyronine levels in cerebrospinal fluid of patients with endogens depression. Acta Endocrinol (Copenh). 1991;124(2):166-72. Cited in PubMed; PMID 1900653.

93. Kishimoto T, Radulovic J, Radulovic M, Lin CR, Schrick C, Hooshmand F, Hermanson O, Rosenfeld MG, Spiess J. Deletion of CRHR2 reveals an anxiolytic role for corticotropin-releasing hormone receptor-2. Nat Genet. 2000;24(4):415–19. Cited in PubMed; PMID 10742109.

94. Köhrle J. Local activation and inactivation of thyroid hormones: the deiodinase family. Mol Cell Endocrinol. 1999;151(1-2):103-19. Cited in PubMed; PMID 10411325.

95. Konkle AT[1], McCarthy MM. Developmental Time Course of Estradiol, Testosterone, and Dihydrotestosterone Levels in Discrete Regions of Male and Female Rat Brain. Endocrinology. 2011;152(1):223–35. doi: 10.1210/en.2010-0607. Cited in PubMed; PMID 21068160.

96. Koolhaas JM, Schuurman T, Wiepkema PR. The organization of intraspecific agonistic behaviour in the rat. Prog Neurobiol. 1980; 15: 247–68. Cited in PubMed; PMID 7005965.

97.   Krzisch M, Temprana SG, Mongiat LA, Armida J, Schmutz V, Virtanen MA, Kocher-Braissant J, Kraftsik R, Vutskits L, Conzelmann KK, Bergami M, Gage FH, Schinder AF, Toni N. Pre-existing astrocytes form functional perisynaptic processes on neurons generated in the adult hippocampus. Brain Struct Funct. 2014 [Epub ahead of print]. Cited in PubMed; PMID 24748560.

98.   Kubala KH, McGinnis MY, Anderson GM, Lumia AR. The effects of an anabolic androgenic steroid and low serotonin on social and non-social behaviors in male rats. Brain Res. 2008;1232:21-9. doi: 10.1016/j.brainres.2008.07.065. Cited in PubMed; PMID 18692488.

99.   Kudryavtseva NN. A sensory contact model for the study of aggressive and submissive behavior in male mice. Aggr Behav. 1991;17(5):285-91. doi: 10.1002/1098-2337(1991)17:5<285::AID-AB2480170505>3.0.CO;2-P.

100.   Kulikov AV, Kozlachkova EY, Maslova GB, Popova NK. Inheritance of predisposition to catalepsy in mice. Behav Genet. 1993;23(4):379-84. Cited in PubMed; PMID 8240217.

101.   Kulikov AV, Osipova DV, Naumenko VS, Popova NK. Association between Tph2 gene polymorphism, brain tryptophan hydroxylase activity and aggressiveness in mouse strains. Genes Brain Behav. 2005;4(8):482-5. Cited in PubMed; PMID 16268992.

102.   Kulikov AV, Osipova DV, Naumenko VS, Terenina E, Mormède P, Popova NK. A pharmacological evidence of positive association between mouse intermale aggression and brain serotonin metabolism. Behav Brain Res. 2012;233(1):113-9. doi: 10.1016/j.bbr.2012.04.031. Cited in PubMed; PMID: 22561036.

103.   Larisch R, Kley K, Nikolaus S, Sitte W, Franz M, Hautzel H, Tress W, Müller HW. Depression and anxiety in different thyroid function states. Horm Metab Res. 2004;36(9):650-3. Cited in PubMed; PMID 15486818.

104.   Lee KJ, Kim SJ, Kim SW, Choi SH, Shin YC, Park SH, Moon BH, Cho E, Lee MS, Chun BG, Shin KH. Chronic mild stress decreases survival, but not proliferation, of new-born cells in adult rat hippocampus. Exp Mol Med. 2006;38:44–54. Cited in PubMed; PMID 16520552.

105.   Lee S, Jeong J, Kwak Y, Park SK. Depression research: where are we now? Mol Brain. 2010;3:8. doi: 10.1186/1756-6606-3-8. Cited in PubMed; PMID 20219105.

106.   Levin-Allerhand JA, Lominska CE, Wang J, Smith JD. 2002 17α-estradiol and 17β-estradiol treatments are effective in lowering cerebral amyloid-β levels in

AβPPSWE transgenic mice. J Alzheimers Dis. 2002; 4: 449–57. Cited in PubMed; PMID 12515896.

107. Liberman A. Depression in Parkinson's disease: a review. Acta Neurol. Scand. 2006;113:1-8. Cited in PubMed; PMID 16367891.

108. López JF, Chalmers DT, Little KY, Watson SJ. A.E. Bennett Research Award. Regulation of serotonin1A, glucocorticoid, and mineralocorticoid receptor in rat and human hippocampus: implications for the neurobiology of depression. Biol Psychiatry. 1998;43(8):547-73. Cited in PubMed; PMID: 9564441.

109. Lu Y, Simpson KL, Weaver KJ, Lin RC. Differential distribution patterns from medial prefrontal cortex and dorsal raphe to the locus coeruleus in rats. Anat Rec (Hoboken). 2012l;295(7):1192-201. doi: 10.1002/ar.22505. Cited in PubMed; PMID: 22674904.

110. Lund TD, Hinds LR, Handa RJ. The androgen 5alpha-dihydrotestosterone and its metabolite 5alpha-androstan-3beta, 17beta-diol inhibit the hypothalamo-pituitary-adrenal response to stress by acting through estrogen receptor beta-expressing neurons in the hypothalamus. J Neurosci. 2006;26(5):1448–56. Cited in PubMed; PMID 16452668.

111. Lund TD, Munson DJ, Haldy ME, Handa RJ. Dihydrotestosterone may inhibit hypathalamo-pituitary-adrenal activity by acting through estrogen receptor in the male mouse. Neurosci. Lett. 2004;365(1):43-7. Cited in PubMed; PMID 15234470.

112. MacLusky NJ, Luine VN, Hajszan T, Leranth C. The 17α and 17β isomers of estradiol both induce rapid spine synapse formation in the CA1 hippocampal subfield of ovariectomized female rats. Endocrinology. 2004; 146: 287–93. Cited in PubMed; PMID 15486220.

113. MacLusky NJ, Hajszan T, Prange-Kiel J, Leranth C. Androgen modulation of hippocampal synaptic plasticity. Neuroscience. – 2006;138(3):957-65. Cited in PubMed; PMID 16488544.

114. Marco Bortolato, Kevin Chen, Sean C Godar, Gao Chen, Weihua Wu, Igor Rebrin, Mollee R Farrell, Anna L Scott, Cara L Wellman, and Jean C Shih. Social deficits and perseverative behaviors, but not overt aggression, in MAO-A pypomorphic mice Neuropsychopharmacology. 2011; 36(13): 2674–2688. doi: 10.1038/npp.2011.157. Cited in PubMed; PMID 21832987.

115. Marino MD, Bourdélat-Parks BN, Cameron Liles L, Weinshenker D. Genetic reduction of noradrenergic function alters social memory and reduces

aggression in mice. Behav Brain Res. 2005;161(2): 197-203. Cited in PubMed; PMID 15922045.

116. Markevich VA, Voronin LL, Mats VN, Tsippel' U. Effect of stimulation of "emotiogenic" brain structures on pyramidal tract responses. Zh Vyssh Nerv Deiat Im I P Pavlova. 1983;33(3):535-42. Cited in PubMed; PMID: 6613346.

117. Maruyama M, Matsumoto H, Fujiwara K, Noguchi J, Kitada C, Fujino M, Inoue K. Prolactin-releasing peptide as a novel stress mediator in the central nervous system. Endocrinology. 2001;142(5): 2032-8. Cited in PubMed; PMID 11316770.

118. Mathews TA, Fedele DE, Coppelli FM, Avila AM, Murphy DL, Andrews AM. Gene dose-dependent alterations in extraneuronal serotonin but not dopamine in mice with reduced serotonin transporter expression. J Neurosci Methods. 2004;140(1-2):169-81. Cited in PubMed; PMID 15589347.

119. Mattay V S, Goldberg T E, Fera F, Hariri A H, Tessitore A, Egan M F, Kolachana B, Callicott J H, and Weinberger D R. Catechol O-methyltransferase val^{158}-met genotype and individual variation in the brain response to amphetamine Proc Natl Acad Sci USA. 2003; 100(10): 6186–6191. doi: 10.1073/pnas.0931309100. Cited in PubMed; PMID12716966. PMC156347.

120. McDevitt RA, Neumaier JF. Regulation of dorsal raphe nucleus function by serotonin autoreceptors: a behavioral perspective. J Chem Neuroanat. 2011;41(4):234-46. doi: 10.1016/j.jchemneu.2011.05.001. Cited in PubMed; PMID 21620956.

121. McEwen B.S. Physiology and neurobiology of stress and adaptation: central role of the brain. Physiol Rev. 2007;87(3):873–904. Cited in PubMed; PMID 12614635.

122. Meloni G., Reedy C.L., Cohen B.M., Carlezon W.A., "Activation of raphe efferents to the medical prefrontal cortex by CRF, correlation with anxiety-like behavior", Biol. Psychiatry, 2008;63(9):832-9. Cited in PubMed; PMID 18061145.

123. Mikics E, Kruk MR, Haller J. Genomic and non-genomic effects of glucocorticoids on aggressive behavior in male rats. Psychoneuroendocrinology. 2004; 29:618–35. Cited in PubMed; PMID 15041085.

124. Millan MJ. The neurobiology and control of anxious states. Prog Neurobiol. 2003;70(2):83-244. Cited in PubMed; PMID 12927745.

125. Moreom X, Jeanningros R, Mazzola-Pomietto P. Chronic effects of triiodothyronine in combination with imipramine on 5-HT transporter, 5-HT (1A) and

5 – HT (2A) receptors in adult rat brain. Neuropsychopharmacology. 2001;24(6):652-62. Cited in PubMed; PMID 11331145.

126.   Morrison SF, Cao WH. Different adrenal sympathetic preganglionic neurons regulate epinephrine and norepinephrine secretion. Am J Physiol Regul Integr Comp Physiol. 2000; 279(5): 1763-75. Cited in PubMed; PMID 11049860.

127.   Mosienko V, Bert B, Beis D, Matthes S, Fink H, Bader M, and Alenina N. Exaggerated aggression and decreased anxiety in mice deficient in brain serotonin. Transl Psychiatry. 2012; 2(5): e122. doi: 10.1038/tp.2012.44 Cited in PubMed; PMCID: PMC3365263. (

128.   Moura E[1], Afonso J, Hein L, Vieira-Coelho MA. Alpha2-adrenoceptor subtypes involved in the regulation of catecholamine release from the adrenal medulla of mice. Br J Pharmacol. 2006:149;1049-58. Cited in PubMed; PMID 17075569.

129.   Moura EG, Moura CC. Regulation of thyrotropin syntheses and secretion. Arq Bras Endocrinol Metabol. 2004;48(1):40-52. Cited in PubMed; PMID 15611817.

130. Murano M, Saitow F, Suzuki H. Modulatory effects of serotonin on glutamatergic synaptic transmission and long-term depression in the deep cerebellar nuclei. Neuroscience. 2011;172:118–28. doi: 10.1016/j. Cited in PubMed; PMID 2096992.

131.   Nemeroff CB. Clinical significal of psychoneuroendocrinology in psychiatry: focus on the thyroid and adrenal. J Clin Psychiatry. 1989;50 Suppl:13-20; discussion 21-2. Cited in PubMed; PMID 2654128.

132.   Neumann ID, Veenema AH, Beiderbeck DI. Aggression and anxiety: social context and neurobiological links. Front Behav Neurosci. 2010;4:12. doi: 10.3389/fnbeh.2010.00012. eCollection 2010. Cited in PubMed; PMID 20407578.

133.   Ni X, Nicholson RC. Steroid hormone mediated regulation of corticotrophin releasing hormone gene expression. Front. Biosci. 2006;11: 2909-17. Cited in PubMed; PMID 16720362.

134.   Nomura M, Durbak L, Chan J, Smithies O, Gustafsson JA, Korach KS et al. Genotype/age interactions on aggressive behavior in gonadally intact estrogen receptor beta knockout (betaERKO) male mice. Horm Behav. 2002; 41(3): 288–96. Cited in PubMed; PMID 11971662.

135.   Nunes-de-Souza V, Nunes-de-Souza R, Rodgers RJ, Canto-de-Souza A. Blockade of 5-HT(2) receptors in the periaqueductal grey matter (PAG) abolishes the anxiolytic-like effect of 5-HT(1A) receptor antagonism in the median raphe nucleus

in mice. Behav Brain Res. 2011;225(2):547-53. doi: 10.1016/j.bbr.2011.07.056. Cited in PubMed; PMID: 21839779.

136.   Olanow CW, Koller WC. An algorithm (decision tree) for the management of Parkinson's disease: treatment guidelines. Neurology. 2011; 50(3): 1-63. Cited in PubMed; PMID 9524552.

137. Oostland M, Sellmeijer J, van Hooft JA. Transient expression of functional serotonin 5-HT3 receptors by glutamatergic granule cells in the early postnatal mouse cerebellum. J Physiol. 2011;589(Pt 20):4837-46. doi: 10.1113/jphysiol.2011.217307. Cited in PubMed; PMID 21878518.

138.   Oquendo MA[1], Mann JJ. The biology of impulsivity and suicidality. Psychiatr Clin North Am. 2000;23(1):11-25. Cited in PubMed; PMID 10729928.

139.   Orsini CA, Kim JH, Knapska E, Maren S. Hippocampal and prefrontal projections to the basal amygdala mediate contextual regulation of fear after extinction. J Neurosci. 2011;31(47):17269-77. doi: 10.1523/JNEUROSCI.4095-11.2011. Cited in PubMed; PMID: 22114293.

140.   Palha JA. Transthyretin as a thyroid hormone carrier: function revisited. Clin Chem Lab Med. 2002;40(12):1292-300. Cited in PubMed; PMID 12553433.

141.   Parker KJ, Schatzberg AF, Lyons DM. Neuroendocrine aspects of hypercortisolism in major depression. Horm Behav. 2003;43(1):60-6. Cited in PubMed; PMID 12614635.

142. Pattij T, Janssen MC, Vanderschuren LJ, Schoffelmeer AN, van Gaalen MM. Involvement of dopamine D1 and D2 receptors in the nucleus accumbens core and shell in inhibitory response control. Psychopharmacology (Berl). 2007;191(3):587-98. Cited in PubMed; PMID16972104.

143. Pavlov KA, Chistiakov DA, Chekhonin VP. Genetic determinants of aggression and impulsivity in humans. J Appl Genet. 2012;53(1):61-82. doi: 10.1007/s13353-011-0069-6. Cited in PubMed; PMID 21994088.

144. Picazo O, Chuc-Meza E, Anaya-Martinez V, Jimenez I, Aceves J, Garcia-Ramirez M. 6-Hydroxydopamine lesion in thalamic reticular nucleus reduces anxiety behaviour in the rat. Behav Brain Res. 2009;197(2):317-22. doi: 10.1016/j.bbr.2008.08.047. Cited in PubMed; PMID 18824199.

145.   Pivonello R, Ferone D, Lombardi G, Colao A, Lamberts SW, Hofland LJ. Novel insights in dopamine receptor physiology. Eur J Endocrinol. 2007;156(1):13-21. Cited in PubMed; PMID 17413183.

146. Popova LD, Vasil`yeva IM. Roles of Central Monoaminergic Systems in the Formation of Different Types of Aggressiveness in Rats. Neurophysiology. 2014;46(3):263-6. http://link.springer.com/article/10.1007%2Fs11062-014-9438-1.

147. Popova NK, Vishnivetskaya GB, Ivanova EA, Skrinskaya JA, Seif I. Altered behavior and alcohol tolerance in transgenic mice lacking MAO A: a comparison with effects of MAO A inhibitor clorgyline. Pharmacol Biochem Behav. 2000;67(4):719-27. Cited in PubMed; PMID 11166062.

148. Popova NK. From gene to aggressive behavior: the role of brain serotonin. Neurosci Behav Physiol. 2008;38(5):471-5. doi: 10.1007/s11055-008-9004-7. Cited in PubMed; PMID 18607754.

149. Posdeyeva EA. Hypothesis affective disorders based on neuroplasticity. A new view at the theory of depression. Psychiatry and psychopharmacology. 2007;43(1):49-52.

150. Prokopova I. Noradrenalin and behavior. Cesk. Fysiol. 2010;59(2): 51-8. Cited in PubMed; PMID 21254660.

151. Przegalinski E, Jaworska L, Konarska R, Gołembiowska K. The role of dopamine in regulation of thyrotropin releasing hormone in the striatum and nucleus accumbens of the rat. Neuropeptides. 1991;19(3):189-95. Cited in PubMed; PMID 1680223.

152. Raison CL, Miller AH. When not enough is too much: the role of insufficient glucocorticoid signaling in the pathophysiology of stress-related disorders. Am J Psychiatry. 2003;160(9):1554–65. Cited in PubMed; PMID 12944327.

153. Raskin K, de Gendt K, Duittoz A, Liere P, Verhoeven G, Tronche F, Mhaouty-Kodja S. Conditional inactivation of androgen receptor gene in the nervous system: effects on male behavioral and neuroendocrine responses. J Neurosci. 2009;29(14):4461-70. doi: 10.1523/JNEUROSCI.0296-09.2009. Cited in PubMed; PMID 19357272.

154. Ressler KJ, Nemeroff CB. Role of serotonergic and noradrenergic systems in the pathophysiology of depression and anxiety disorders. Depress Anxiety. 2000;12(1):2-19. Cited in PubMed; PMID 11098410.

155. Reul JM, Holsboer F. Corticotropin-releasing factor receptors 1 and 2 in anxiety and depression. Curr Opin Pharmacol. 2002;2(1):23-33. Cited in PubMed; PMID 11786305.

156. Reznikov AG, Nosenko ND, Tarasenko LV, Sinitsyn PV, Lymareva AA. Prenatal dexamethasone prevents early and long-lasting neuroendocrine and behavioral effects of maternal stress on male offspring. Fiziol Zh. 2008;54(5):28-39.

Cited in PubMed; PMID 19058510.

157. Reznikov AG. Mechanisms of development of functional pathology of reproduction and adaptation in early ontogenesis. Zh AMN of Ukraine. 1998;4(2):216-33.

158. Rhodes MC, Seidler FJ, Abdel-Rahman A, Tate CA, Nyska A, Rincavage HL, Slotkin TA. Terbutaline is a developmental neurotoxicant: effects on neuroproteins and morphology in cerebellum, hippocampus, and somatosensory cortex. J Pharmacol Exp Ther. 2004;308(2):529-37. Cited in PubMed; PMID 14610225.

159. Risold PY, Swanson LW. Connections of the rat lateral septal complex Brain Res Brain Res Rev. 1997;24(2-4):115–95. Cited in PubMed; PMID 9385454.

160. Robertson SD, Plummer NW, de Marchena J, Jensen P. Developmental origins of central norepinephrine neuron diversity. Nat Neurosci. 2013;16(8):1016-23. doi: 10.1038/nn.3458. Cited in PubMed; PMID: 23852112.

161. Roselli CE, Liu M, Hurn PD. Brain aromatization. Classical roles and new perspectives. Semin. Reprod. Med. 2009;27(3):207-27.

162. Roth BL. The multiplicity of serotonin receptors: uselessly diverse molecules or an embarrassment of riches? /B.L. Roth // Neuroscientist. – 2000. – Vol. 6, N 4. – P. 252-262.

163. Samuels ER, Zabadi ES. Functional neuroanatomy of the noradrenergic locus coeruleus: its roles in the regulation of arousal and autonomic function. Part I: Principles of functional organization. Curr. Neuropharmacol. 2008;6(3): 235-53. doi: 10.2174/157015908785777229. Cited in PubMed; PMID 9506723.

164. Sapronov NS, Maslova OO Neurophysiological effects of thyroid hormons. Psychopharmacol Biol Narcol. 2007;7(2): 1533-41.

165. Sato T, Matsumoto T, Kawano H, Watanabe T, Uematsu Y, Sekine K, Fukuda T, Aihara K, Krust A, Yamada T, Nakamichi Y, Yamamoto Y, Nakamura T, Yoshimura K, Yoshizawa T, Metzger D, Chambon P, Kato S. Brain masculinization requires androgen receptor function. Proc Natl Acad Sci U S A. 2004;101(6):1673-8. Cited in PubMed; PMID 14747651.

166. Schlumpf M, Lichtensteiger W, Langemann H, Waser PG, Hefti F. A fluorometric micromethod for the simultaneous determination of serotonin, noradrenaline and dopamine in milligram amounts of brain tissue. Biochem Pharmacol. 1974;23(17):2437-46. Cited in PubMed; PMID 4429570.

167. Scordalakes EM, Rissman EF. Aggression and arginine vasopressin immunoreactivity regulation by androgen receptor and estrogen receptor alpha. Genes Brain Behav. 2004; 3(1):20–6. Cited in PubMed; PMID 14960012.

168.   Semenenya I.N. Functional meaning of thy roid gland. Usp Fiziol Nauk. 2004;35(2):41-56.

169.   Senn V, Wolff SB, Herry C, Grenier F, Ehrlich I, Gründemann J, Fadok JP, Müller C, Letzkus JJ, Lüthi A. Long-range connectivity defines behavioral specificity of amygdala neurons. Neuron. 2014;81(2):428-37. doi: 10.1016/j.neuron.2013.11.006. Cited in PubMed; PMID: 24462103.

170.   Shabanov PD, Lebedev AA. Neurochemical mechanisms of the nucleus accumbens realizing the reinforcing effects of self–stimulation on the lateral hypothalamus. Med. Acad. Journal. 2012; 12(2): 68-76.

171.   Sharara-Chami RI, Joachim M, Pacak K, Majzoub JA.Glucocorticoid treatment--effect on adrenal medullary catecholamine production. Shock. 2010; 33(2): 213-17. doi: 10.1097/SHK.0b013e3181af0633. Cited in PubMed; PMID 19503019.

172. Shih JC, Chen K, Ridd MJ. Monoamine oxidase: from genes to behavior. Annu Rev Neurosci. 1999;22:197–217. Cited in PubMed; PMID 10202537.

173.   Shimizu T, Yokotani K. Brain cyclooxygenase and prostanoid TP receptors are involved in centrally administered epibatidine-induced secretion of noradrenaline and adrenaline from the adrenal medulla in rats. Eur J Pharmacol. 2009;606(1-3):77-83. Cited in PubMed; PMID 19374850.

174.   Shin LM, Liberzon I. The neurocircuitry of fear, stress, and anxiety disorders. Neuropsychopharmacology. 2010;35(1):169-91. doi: 10.1038/npp.2009.83. Cited in PubMed; PMID: 19625997.

175.   Sintzel F, Mallaret M, Bougerol T. Potentializing of tricyclics and serotoninergics by thyroid hormones in resistant depressive disorders. Encephale. 2004;30(3):267-75. Cited in PubMed; PMID 15235525.

176.   Sotnikova TD, Budygin EA, Jones SR, Dykstra LA, Caron MG, Gainetdinov RR. Dopamine transporter-dependent and -independent actions of trace amine beta-phenylethylamine. J Neurochem. 2004;91(2):362-73. Cited in PubMed; PMID 15447669.

177.   Stanley JA, Aruldhas MM, Yuvaraju PB, Banu SK, Anbalagan J, Neelamohan R, Annapoorna K, Jayaraman G. Is gender difference in postnatal thyroid growth associated with specific expression patterns of androgen and estrogen receptors? Steroids. 2010;75(13-14):1058-66. doi: 10.1016/j.steroids.2010.06.009. Cited in PubMed; PMID 20670640.

178.   Steimer T. The biology of fear- and anxiety-related behaviors. Dialogues Clin Neurosci. 2002;4(3):231-49. Cited in PubMed; PMID: 22033741.

179. Steinlin M. Cerebellar disorders in childhood: cognitive problems. Cerebellum. 2008;7(4):607-10. doi: 10.1007/s12311-008-0083-3. Cited in PubMed; PMID 19057977.

180. Stone EA, Quartermain D, Lin Y, Lehmann ML. Central alpha1-adrenergic system in behavioral activity and depression. Biochem Pharmacol. 2007;73(8):1063-75. Cited in PubMed; PMID: 17097068.

181. Stroth N, Eiden LE. Stress hormone synthesis in mouse hypothalamus and adrenal gland triggered by restraint is dependent on pituitary adenylate cyclase-activating polypeptide signaling. Neuroscience. 2010 Feb 17;165(4):1025-30. doi: 10.1016/j.neuroscience.2009.11.023. Epub 2009 Nov 18.

182. Stuber GD, Sparta DR, Stamatakis AM, van Leeuvan WA, Hardjoprajitno JE, Cho S, Tye KM, Kempadoo KA, Zhang F, Deisseroth K, Bonci A. Excitatory transmission from the amygdala to nucleus accumbens facilitates reward seeking. Nature. 2011; 475(7356): 377-80. doi: 10.1038/nature10194. Cited in PubMed; PMID 21716290.

183. Summers CH, Winberg S. Interactions between the neural regulation of stress and aggression. J Exp Biol. 2006;209(Pt 23):4581-9. Cited in PubMed; PMID 17114393.

184. Surget A, Saxe M, Leman S, Ibarguen-Vargas Y, Chalon S, Griebel G, Hen R, Belzung C. Drug-dependent requirement of hippocampal neurogenesis in a model of depression and of antidepressant reversal. Biol Psychiatry. 2008;64:293–301. doi: 10.1016/j.biopsych.2008.02.022. Cited in PubMed; PMID 18406399.

185. Susman EJ, Inoff-Germain G, Nottelmann ED, Loriaux DL, Cutler GB, Chrousos GP. Hormones, emotional dispositions, and aggressive attributes in young adolescents. Child Dev. 1987;58:1114–34. Cited in PubMed; PMID 3608660.

186. Swaab DF, Bao AM, Lucassen PJ. The stress system in the human brain in depression and neurodegeneration. Ageing Res Rev. 2005;4(2): 141-94. Cited in PubMed; PMID 15996533.

187. Tahboub R, Arofan BM. Sex steroids and the thyroid. Best Pract Res Clin Endocrinol Metab. 2009;23(6):769-80. doi: 10.1016/j.beem.2009.06.005. Cited in PubMed; PMID 19942152.

188. Tanaka M, Yoshida M, Emoto H, Ishii H.Noradrenaline systems in the hypothalamus, amygdala and locus coeruleus are involved in the provocation of anxiety: basic studies. Eur J Pharmacol. 2000;405(1-3):397-406. Cited in PubMed; PMID 11033344.

189. Tepperman J, Tepperman HM. Metabolic and endocrine physiology an

introductory text. 5th ed. Chicago-London, Year book Medical Publishers, 1987 p.177-220.

190. Thompson BL, Stanwood GD. Pleiotropic effects of neurotransmission during development: modulators of modularity. J Autism Dev Disord. 2009;39(2):260-8. doi: 10.1007/s10803-008-0624-0. Cited in PubMed; PMID 18648918.

191. Thompson CC, Potter GB. Thyroid hormones action in neural development Cereb Cortex. 2000;10(10):939-45. Cited in PubMed; PMID 11007544.

192. Toran-Allerand CD, Tinnikov AA, Singh RJ, Nethrapalli IS. 17α-Estradiol: A Brain-Active Estrogen? Endocrinology. 2005; 146(9):3843-50. Cited in PubMed; PMID 15947006.

193. Tse WS, Bond AJ. Difference in serotonergic and noradrenergic regulation of human social behaviours.Psychopharmacology (Berl). 2002; 159(2):216-21. Cited in PubMed; PMID 11862352.

194. Tse WS, Bond AJ. Reboxetine promotes social bonding in healthy volunteers. J Psychopharmacol. 2003;17(2):189-95. Cited in PubMed; PMID 12870566.

195. Tulogdi A, Toth M, Halasz J, Mikics E, Fuzesi T, Haller J. Brain mechanisms involved in predatory aggression are activated in a laboratory model of violent intra-specific aggression. Eur J Neurosci. 2010;32(10):1744-53. doi: 10.1111/j.1460-9568.2010.07429.x. Cited in PubMed; PMID: 21039962.

196. Vaccari A, Caviglia A, Sparatore A, Biassoni R. Gonadal influences on the sexual differentiation of monoamine oxidase type A and B activities in the rat brain. J Neurochem. 1981;37(3):640-8. Cited in PubMed; PMID 7276946.

197. van Bokhoven I., van Goozen SH, van Engeland H, Schaal B, Arseneault L, Seguin JR et al. Salivary testosterone and aggression, delinquency, and social dominance in a population-based longitudinal study of adolescent males. Horm Behav. 2006;50:118–25. Cited in PubMed; PMID 16631757.

198. van der Vegt BJ, Lieuwes N, Cremers TI, de Boer SF, Koolhaas JM. Cerebrospinal fluid monoamine and metabolite concentrations and aggression in rats. Horm Behav. 2003;44(3):199-208. Cited in PubMed; PMID 14609542.

199. van Gaalen MM, Brueggeman RJ, Bronius PF, Schoffelmeer AN, Vanderschuren LJ. Behavioral disinhibition requires dopamine receptor activation. Psychopharmacology (Berl). 2006;187(1):73-85. Cited in PubMed; PMID 16767417.

200. van Wingen GA, Ossewaarde L, Bäckström T, Hermans EJ, Fernandez G. Gonadal hormone regulation of the emotion circuitry in humans. Neuroscience. 2011;191:38–45.

201. Vasylyeva IM, Popova LD. Optimization of control choice when using sensory contact model. Experimental and clinical medicine. 2010; 48(3):37-40.

202. Veenema AH, Sijtsma B, Koolhaas JM, de Kloet ER. The stress response to sensory contact in mice: genotype effect of the stimulus animal. Psychoneuroendocrinology. 2005;30:550–7. Cited in PubMed; PMID 15808924.

203. Viau V, Meaney MJ. Testosterone-dependent variations in plasma and intrapituitary corticosteroid binding globulin and stress hypothalamic-pituitary-adrenal activity in the male rat. J Endocrinol. 2004;181:223–31. Cited in PubMed; PMID 15128271.

204. Vitiello B, Stoff DM. Subtypes of aggression and their relevance to child psychiatry. J Am Acad Child Adolesc Psychiatry. 1997;36(3):307–15. 10.1097/00004583-199703000-00008. Cited in PubMed; PMID 9055510.

205. Vollmer RR. Selective neural regulation of epinephrine and norepinephrine cells in the adrenal medulla -- cardiovascular implications. Clin Exp Hypertens. 1996;18(6):731-51. Cited in PubMed; PMID 8842561.

206. Walker DL, Toufexis DJ, Davis M. Role of the bed nucleus of the stria terminalis versus the amygdala in fear, stress, and anxiety. Eur J Pharmacol. 2003 Feb 28;463(1-3):199-216. Cited in PubMed; PMID: 12600711.

207. Warden MR, Selimbeyoglu A, Mirzabekov JJ, Lo M, Thompson KR, Kim SY, Adhikari A, Tye KM, Frank LM, Deisseroth K. A prefrontal cortex-brainstem neuronal projection that controls response to behavioural challenge. Nature. 2012;492(7429):428-32. doi: 10.1038/nature11617. Cited in PubMed; PMID: 23160494.

208. Weiser MJ, Goel N, Sandau US, Bale TL, Handa RJ. Androgen regulation of corticotropin-releasing hormone receptor 2 (CRHR2) mRNA expression and receptor binding in the rat brain. Exp Neurol. 2008; 214(1):62–8. doi: 10.1016/j.expneurol.2008.07.013. Cited in PubMed; PMID 18706413.

209. Wingfield JC, Lynn S, Soma KK. Avoiding the 'costs' of testosterone: ecological bases of hormone-behavior interactions. Brain Behav. 2001;57:239–51.Cited in PubMed; PMID 11641561.

210. Winstanley CA, Theobald DE, Dalley JW, Cardinal RN, Robbins TW. Double dissociation between serotonergic and dopaminergic modulation of medial prefrontal and orbitofrontal cortex during a test of impulsive choice. Cereb Cortex. 2006;16(1):106–14. Cited in PubMed; PMID 15829733.

211. Wiskott L, Rasch MJ, Kempermann G. A functional hypothesis for adult hippocampal neurogenesis: avoidance of catastrophic interference in the dentate gyrus. Hippocampus. 2006;16(3):329-43. Cited in PubMed; PMID: 16435309.

212. Witte AV, Flöel A, Stein P, Savli M, Mien LK, Wadsak W, Spindelegger C, Moser U, Fink M, Hahn A, Mitterhauser M, Kletter K, Kasper S, Lanzenberger R. Aggression is related to frontal serotonin-1A receptor distribution as revealed by PET in healthy subjects. Hum Brain Mapp. 2009;30(8):2558-70. doi: 10.1002/hbm.20687. Cited in PubMed; PMID 19086022.

213. Woodward DJ, Moises HC, Waterhouse BD, Yeh HH, Cheun JE. The cerebellar norepinephrine system: inhibition, modulation, and gating. Prog Brain Res. 1991;88:331-41. Cited in PubMed; PMID 1687621.

214. Yen P.M. Physiological and molecular basis of thyroid hormone action. Physiol Rev. 2001;81(3):1097-142. Cited in PubMed; PMID 11427693.

215. Yuen EY, Jiang Q, Chen P, Gu Z, Feng J, Yan Z. Serotonin 5-HT1A receptors regulate NMDA receptor channels through a microtubule-dependent mechanism. J Neurosci. 2005;25(23):5488-501. Cited in PubMed; PMID 15944377.

216. Zaichenco MI, Mikhailova NG, RaigorodskiiYuV. Neuron activity in the prefrontal cortex of the brain in rats with different typological characteristics in conditions of emotional stimulation. Neurosci Behav Physiol. 2001;31(3):299-304. Cited in PubMed; PMID 11430574.

217. Zapadnyuk YP, Zapadnyuk VY, Zakharyya EA, Zapadnyuk BV. Laboratory animals. Kyiv: High school; 1983.

218. Zhang X, Beaulieu JM, Sotnikova TD, Gainetdinov RR, Caron MG. Tryptophan hydroxylase-2 controls brain serotonin synthesis. Science. 2004;305(5681):217. Cited in PubMed; PMID 15247473.

219. Zuloaga DG, Puts DA, Jordan CL and Breedlove SM. The Role of Androgen Receptors in the Masculinization of Brain and Behavior: What we've learned from the Testicular Feminization Mutation. Horm Behav. 2008;53(5):613–26. doi: 10.1016/j.yhbeh.2008.01.013. Cited in PubMed; PMID 18374335.

220. Zusso M, Debetto P, Guidolin D, Barbierato M, Manev H, Giusti P. Fluoxetine-induced proliferation and differentiation of neural progenitor cells isolated from rat postnatal cerebellum. Biochem Pharmacol. 2008;76(3):391-403. doi: 10.1016/j.bcp.2008.05.014. Cited in PubMed; PMID 18573488.

Printed by Books on Demand GmbH, Norderstedt / Germany